Challenges in Fine Coal Processing, Dewatering, and Disposal

Challenges in Fine Coal Processing, Dewatering, and Disposal

Editor

Neeraj Singh

Challenges in Fine Coal Processing, Dewatering, and Disposal

Edited by **Neeraj Singh**

Printed in 2017

ISBN: 978-1-68117-475-4

Library of Congress Control Number: 2015936592

© 2016 by
SCITUS Academics LLC,
616, Corporate Way, Suite 2, 4766,
Valley Cottage, NY 10989

www.scitusacademics.com

Contents

Preface

Coal mining and preparation have had a long history in the United States and the world, serving as the engine of growth for many industries. Today, new sources of energy, increased environmental awareness and more stringent regulations from the U.S. Environmental Protection Agency and other organizations are changing the way coal is found, extracted and used. As a result, fine coal cleaning, dewatering and refuse disposal are now at a major crossroads. The increased level of fines, and near-density material in the inferior seams being mined today, necessitates the development of more efficient fine coal cleaning devices. This in turn requires improvements in traditional dewatering techniques to address the need for acceptable moisture levels in plant products. Moreover, the larger volume of fine refuse being generated, coupled with harsher disposal regulations, require upgraded treatment options.

Editor

Chapter 1

Influence of Fly Ash, Bottom Ash, and Light Expanded Clay Aggregate on Concrete

S. Sivakumar[1] and B. Kameshwari[2]

[1]NPR College of Engineering and Technology, NPR Nagar, Natham, Dindigul, Tamil Nadu 624401, India

[2]Ratnavel Subramaniam College of Engineering and Technology, Dindigul, Tamil Nadu 624005, India

ABSTRACT

Invention of new methods in strengthening concrete is under work for decades. Developing countries like India use the extensive reinforced construction works materials such as fly ash and bottom ash and other ingredients in RCC construction. In the construction industry, major attention has been devoted to the use of fly ash and bottom ash as cement and fine aggregate replacements. In addition, light expanded clay aggregate has been introduced instead of coarse aggregate to make concrete have light weight. This paper presents the results of a real-time work carried out to form light weight concrete made with fly ash, bottom ash, and light expanded clay aggregate as mineral admixtures. Experimental investigation on concrete mix M_{20} is done by replacement of cement with fly ash, fine aggregate with bottom ash, and coarse aggregate with light expanded clay aggregate at the rates of 5%, 10%, 15%, 20%, 25%, 30%, and 35% in each mix and their compressive strength and split tensile strength of concrete were discussed for 7, 28, and 56 days and flexural strength has been discussed for 7, 28, and 56 days depending on the optimum dosage of replacement in compressive strength and split tensile strength of concrete.

INTRODUCTION

High performance concrete indicates an exceptional form of concrete endowed with astonishing proficiency and strength essentials which are unequipped with periodical assessment on a regular basis by way of traditional materials and standard mixing, placing, and curing techniques [1]. Ordinary Portland cement (OPC) has grabbed an unenviable and undefeatable position as a significant material in the generation of concrete and meticulously releases its designed obligation as an extraordinary binder to join all the gathered materials. With the purpose of attaining, there is a dire need of the burning of mammoth measure of fuel and rot of limestone [2]. A few grades of ordinary Portland cement (OPC) are accessible tailor made to suit the particular nation code categorization. In this respect, Bureau of Indian Standard (BIS) exquisitely does the capacity of categorizing three separate grades of OPC, for example, 33, 43, and 53, which have chronically been widely utilized in construction industry [3]. The strength, sturdiness, and different attributes of concrete rely on the properties of its ingredients, the proportion of the mix, the strategy of compaction, and different controls amid placing, compaction, and curing [4]. Concrete containing wastes can help construction manageable quality and contribute to the advancement of the civil engineering region by employing industrial waste, minimizing the utilization of natural assets, and producing more effective materials [5]. The Portland cement concrete resorts to the employment of fly ash when the loss-on-ignition (LOI) qualities fall inside the area of 6%. The fly ash is home to the crystalline and amorphous components together with unburnt carbon. It grasps differing measures of unburnt carbon, which is prone to reach the tune of 17% [6]. Fly ash is regularly alluded to as pond ash and over the long haul the water is permitted to drain away. Both techniques viably prompt to dumping of the fly ash in landfills on open land. The chemical composition of fly ash remains changes relying on the type of coal utilized as a part of combustion, combustion conditions, and evacuation productivity of air contamination control device [7]. The effect of fly ash substance and substitution of trampled sand stone total with concrete squandrs and marble squanders employed prefabricated concrete interlocking squares [8]. With an eye on the power of concrete edifices, modern concrete methodology set down extraordinary steps to chop down summit and differential temperatures by deploying materials with the minimized level of release of heat to steer clear of or then again bring down thermal splitting prompting the prevention of the decomposition of the concrete [9]. Production of concrete is done in exceedingly high and imperceptibly low temperatures of concrete to understand the workability and compressive quality [10]. The statistical model and the kinetic property of flexural, breaking tensile furthermore modulus of versatility as per the compressive stability stemmed from the unwarranted coefficient of correlation [11]. Concrete generated out of minute total and superior void ratio is known to be enriched with a brilliant expertise to exile the materials [12]. In India, the power division focused around coal based thermal power stations produces a colossal quantity of fly ash assessed around 11 crore tonnes every annum. The consumption of fly ash is assessed to be around 30% for the purpose of various engineering properties essentials [13]. Ignition of coal to deliver power in a boiler yields around 80% of the unburned material or ash, which is entrained in the flue gas and is entrapped and reclaimed in the shape of fly ash. The residual 20% of the ash helps dry base ash [14]. At the point when pulverized coal is blazed in a dry bottom boiler, around 80 to 90% of the unburned material or ash is entrained in the flue gas and is trapped and recovered as fly ash. The residual 10–20% of the ash is indicated to dry bottom ash, sand size, material which is assembled in water-filled containers at the base of the furnace [15]. Coal bottom ash in concrete is created by the method of fractional, almost-aggregate, and total substitution of fine aggregates in concrete [16]. On the other hand, lightweight concrete is awkward to make a case belong to a unique category material. However, LWC (light weight concrete) has clear edges, and the plunge in the total expense created by the lower dead loads is constantly overshot by the raised production outlay [17]. As a matter of fact, lightweight concrete has surfaced as the agreeable favorite as against the standard concrete in the perspective of a multitude of unrivaled attributes. The dip in dead weight generally brings about cutbacks in production outlay [18]. The self-compacting normal weight aggregate concrete (SCNC) is to be the favorite for the purpose of development. The surge in the construction expense of SCLC fares positively with that for SCNC [19]. Lightweight aggregate concrete deadweight is assessed to be around 15%~30% lighter than standard concrete, which sufficiently fulfills the mechanical attributes that roadway support requires on the specified density degree [20]. Rising utilization of light weight concrete (LWC) brought the requirement for the artificial lightweight total production, which may be accomplished by cold bonding assembling methodology. Production of artificial fly lightweight aggregates with cold bonding process needs much less energy consumption when contrasted with sintering [21]. Lightweight concrete made with natural or artificial lightweight aggregates is accessible in numerous parts of the world. It can be utilized as a part of creating concrete in an extensive variety of unit weights and suitable qualities for different applications [22]. Lightweight aggregates concrete livens up its potency to thwart nearby harm activated by ballistic loading. Lower modulus of flexibility and higher tensile strain limit outfits lightweight concrete opposite standard weight concrete with superior impact resistance [23]. Light concrete material is more and more prescribed by the builders to reach a supportable improvement due to its great strength and thermal properties [24]. The adhesive strength is accomplished from solidity in the binder and interlocking traits of the aggregates, which are constantly focused around angularity, levelness, and extension [25]. Light expanded clay aggregate (LECA), generally, includes tiny, lightweight, bloated particles of burnt clay The hundreds and thousands of tiny, air-filled depressions successfully empower LECA with its sterling strength and thermal insulation qualities. The average water absorption of LECA total (0–25 mm) is thought to associate with 18 percent of volume in saturated status amid the time of 3 days. The ordinary Portland cement (OPC) is partially substituted by the fly ash, fine aggregate interchanged by bottom ash, and coarse aggregate supplanted by light expanded clay aggregate (LECA) by weights of 5%, 10%, 15%, 20%, 25%, 30%,

Influence of Fly Ash, Bottom Ash, and Light Expanded Clay...

5

and 35%, separately. The compressive strength, split tensile strength, and flexural strength are successfully assessed by means of determined input values in concurrent investigation.

EXPERIMENTAL PROGRAM

The objective of the work is to evaluate the compressive strength (CS), split tensile strength (STS), and flexural strength (FS) of the concrete. In this concrete mix, ordinary Portland cement ($OPC_{43grade}$) is replaced by fly ash, the fine aggregate is replaced by bottom ash, and the coarse aggregate is replaced by light expanded clay aggregate (LECA) by weights of 5%, 10%, 15%, 20%, 25%, 30%, and 35%, respectively. For increasing the strength in cement, these materials are to be added. In the experimental investigation, the concrete cube or cylinder is used to analyze the properties of the concrete with all materials. Each weight (5%, 10%, 15%, 20%, 25%, 30%, or 35%) of a material conducted the test on 7 days, 28 days, and 56 days. The parameters involved in evaluating the performance of concrete are compressive strength (CS), split tensile strength (STS), and flexural strength (FS) that are attained from real time experiments. Then finding the flexural strength has been discussed for 7, 28, and 56 days depending on the load for the optimum dosage of replacement in compressive strength and split tensile strength of concrete.

Materials Used

Names of materials used in this research and their performance are listed in this section. The resources are ordinary Portland cement, fly ash, bottom ash, fine aggregate, coarse aggregate and light expanded clay aggregate (LECA).

Ordinary Portland Cement

Ordinary Portland cement is the basic form of cement where 95% of it is clinker and 5% is gypsum which is added as an additive to enhance the setting time of the cement to a workable 30 minutes odd or so. Gypsum controls initial setting time of the cement. If gypsum is not added, cement would be set as soon as water is added in cement.

Different grades (33, 43, and 53) of OPC have been classified by the Bureau of Indian Standards (BIS). It is manufactured in larger quantities when compared with the other types of cement and it is admirably suited for use in general concrete construction where there is no exposure to sulphates in the soil or in ground water. In this research, the cement ($OPC_{43grade}$) that has a specific gravity of 3.15 and initial and final setting times of the cement of 50 minutes and 450 minutes has been used.

Fly Ash

The most common type of coal-burning furnace in the electric utility industry, about 80% of the unburned material or ash, is entrained in the flue gas and is captured and recovered as fly ash. Fly ash was collected from Thoothukudi Thermal Power Plant, Tamil Nadu, India The increasing scarcity of raw materials and the urgent need to protect the environment against pollution have accentuated the significance of developing new building materials based on industrial waste generated from coal fired thermal power station which is creating unmanageable disposal problems due to its potential to pollute the environment. As the cost of disposing of fly ash continues to rise, strategies for the recycling of fly ash are environmentally and economically critical. For the source materials the two emerging areas for the recycling of coal fly ash are used as shown in Figure 1(a).

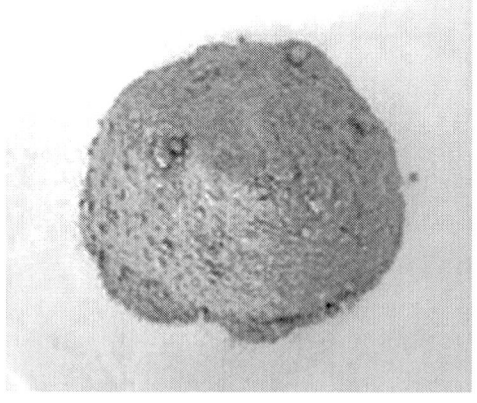

(a)

(b)

(c)

Figure 1: Materials (a) Fly ash, (b) Bottom ash, (c) LECA.

Bottom Ash

The remaining 20% of the unburned material is collected at the bottom of the combustion chamber in a water-filled hopper and is removed by means of high-pressure water jets to a decanting basin for dewatering and is recovered as bottom ash as shown in Figure 1(b). Coal bottom ash was obtained from Thoothukudi Thermal Power Plant, Tamil Nadu, India The fly ash was obtained directly from the bottom of the electrostatic precipitator into a sack because of its powdery and dusty nature, while the coal bottom ash is transported from the bottom of the boiler to an ash pond as liquid slurry where the sample was collected. Bottom ash is lighter and more brittle and it is dark gray material with a grain size similar to that of sand.

Fine Aggregate

According to the Indian standards natural sand is a form of silica (SiO_2) that has maximum particle size of 4.75 mm and it was used as fine aggregate. The minimum particle size of fine aggregate is 0.075 mm. It is formed by decomposition of sand stones due to various weathering actions. Fine aggregate prevents shrinkage of the mortar and concrete. The specific gravity and fineness modulus of coarse aggregate were 2.67 and 2.3.

Fine aggregate is an inert or chemically inactive material, most of which passes through a 4.75 mm IS sieve and contains no more than 5 percent of coarser material. It may be classified as follows:

- Natural sand: fine aggregate that results from the natural disintegration of rocks and has been deposited by streams or glacial agencies;
- Crushed stone sand: fine aggregate produced by crushing hard stone;
- Crushed gravel sand: fine aggregate produced by crushing natural gravel.

It reduces the porosity of the final mass and considerably increases its strength. Usually, natural river sand is used as a fine aggregate. However, at places, where natural sand is not available economically, finely crushed stone may be used as a fine aggregate.

Coarse Aggregate

Coarse aggregate consists of naturally occurring materials such as gravel, or it results from the crushing of parent rock, to include natural rock, slags, expanded clays and shales (lightweight aggregates), and other approved inert materials with similar characteristics, having hard, strong, and durable particles, conforming to the specific requirements of this section.

According to the Indian standards, crushed angular aggregate passes through 20 mm IS sieve and entirely retains 10 mm IS sieve. The specific gravity and fineness modulus of coarse aggregate were 2.60 and 5.95.

Light Expanded Clay Aggregate (LECA)

LECA is shown in Figure 1(c) it has strong resistance against alkaline and acidic substances and pH of nearly 7 makes it neutral in chemical post reaction with concrete. Lightness, insulating, durability, nondecomposability, structural stability, and chemical neutrality features are collected in LECA as the best light weight aggregate for flooring and roofing. The size of the aggregate is 10 mm and the maximum density is less than or equal to 480 Kg/m^3. LECA consists of small, strong, light weight, and thermally insulating particles of burnt clay. LECA which is an environment-friendly and entirely natural product is indestructible, noncombustible, and impervious to attack by dry-rot, wet-rot, and insects. Lightweight concrete is generally classified into two types: aerated concrete (or foamed concrete) and lightweight aggregate concrete. The aerated concrete has very light weight and low thermal conductivity. However, an autoclaving process is essential to obtain a certain level of strength, which requires a special manufacturing plant and consumes very high energy. In contrast, lightweight aggregate concrete, which is manufactured without an autoclaving process, has higher strength but shows higher density and lower thermal conductivity of the concrete.

Conplast Admixture SP430 (G)

Conplast SP430 (G) is used where a high degree of workability and its retention are required when delays in transportation or placing are likely or when high ambient temperatures cause rapid slump loss. It facilitates production of high quality concrete. Conplast SP430 (G) complies with the fact that it has been specially formulated to give high water reductions up to 25% without loss of workability or to produce high quality concrete of reduced permeability. Cohesion is improved due to dispersion of cement particles, thus minimizing segregation and improving surface finish. The optimum dosage is best determined by site trials with the concrete mix which enables the effects of workability, strength gain, or cement reduction to be measured. This type of ingredients is added in concrete to give it certain improved qualities or to change different physical properties in its fresh and hardened stages. Optimum dosage range of cement is 0.6–1.5 liters/100 kg. The addition of an admixture may improve the concrete with respect to its strength, hardness, workability, water resisting power, and so forth.

Structural Specification of Beam

The structural specification of the beam is having top reinforcement diameter of 8 mm, bottom reinforcement diameter of 12 mm, and 6 mm stirrups (Figure 2). Total length of beam utilized for deflection is 1 meter. This specification is used in a concrete structure, and the whole process is done in the specification of a concrete.

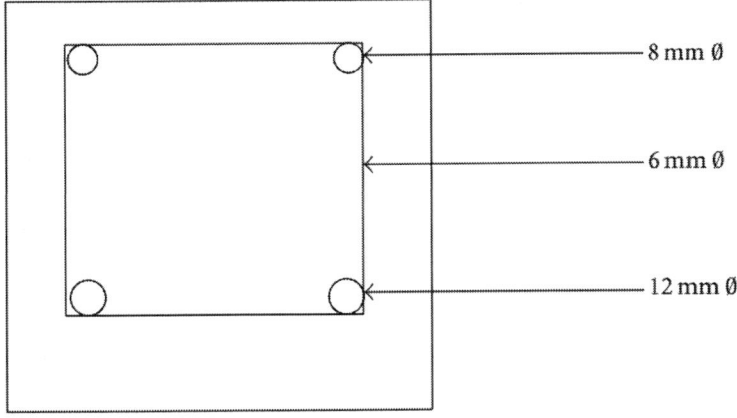

Figure 2: Cross section of beam.

Structural Light Weight Concrete

The concrete is made with a light weight coarse aggregate. Light weight aggregates generally require wetting prior to use to achieve a high degree of saturation. The primary use of structural light weight concrete is to reduce the dead load of a concrete structure. In ordinary concrete different gradation of aggregates affects the required amount of water. Addition of some fine aggregates results in an increase in required amount of water. This increase of water reduces concrete strength unless the amount of cement in the same time increases. Amount of coarse aggregate and its biggest size depend on the required workability of concrete mixture. Also in light weight concrete, this result exists among the gradation, requested amount of water, and obtained concrete strength, but there are other factors that must be paid some attention. In most light weight aggregates as the size of aggregate increases the strength and bulk density of the aggregate decrease. Using very big size light weight aggregate with a lower strength results in a lower strength of the light weight concrete; therefore, biggest size of the light weight aggregate must be limited to 25 mm at most.

METHODOLOGY

Concrete mix proportion for M_{20} grade was obtained based on the guidelines according to Indian standard specifications (IS: 456-2000 and IS: 10262-1982). In this study, experimental investigation on concrete mix M_{20} is done by replacement of cement with fly ash, fine aggregate with bottom ash, and coarse aggregate with light expanded clay aggregate (LECA) at the rates of 5%, 10%, 15%, 20%, 25%, 30%, and 35%, respectively. For increasing the strength in cement, these materials are to be added. In the experimental investigation, the concrete cube or cylinder is used to analyze the properties of the OPC with all materials. Their compressive strength and split tensile strength of concrete have been discussed for 7 days, 28 days, 56 days and flexural strength of beam has been discussed for 7, 28, and 56 days depending on the optimum dosage of replacement in compressive strength and split tensile strength of concrete. Generally, fly ash and bottom ash have similar physical and chemical properties as compared to ordinary Portland cement (OPC) and fine aggregate, and there are not a lot of deviations for replacing each other. In this scenario, light expanded clay aggregate (LECA) has been replaced with coarse aggregate by its volume basis because density of each material is not the same as that of the other material and it is not possible to replace it by its weight basis. To increase the workability of concrete super plasticizer was added.

The ratio of concrete mix M_{20} grade was 1:1.42:3.3. Controlled concrete of M_{20} grade was made with 0% replacement of fly ash, bottom ash, and light expanded clay aggregate (LECA) in each mix and their compressive strength and split tensile strength of concrete have been discussed for 7, 28, and 56 days and flexural strength of concrete has been discussed for 7, 28, and 56 days. In this connection replacement of cement with fly ash, fine aggregate with bottom ash, and coarse aggregate with light expanded clay aggregate (LECA) at the rates of 5%, 10%, 15%, 20%, 25%, 30%, and 35% in each mix was conducted and their compressive strength and split tensile strength of concrete were discussed for 7 days, 28, days, 56 days and flexural strength of beam for 7, 28 and 56 days depends on the optimum dosage of replacement in compressive strength and split tensile strength of concrete.

Water absorption of light weight aggregate with too much pores is much more than ordinary aggregates (river aggregates). Determination of amount of water absorption in these kinds of aggregates is difficult because of the varying amounts of absorbed water. LECA aggregate produced rotary kiln, and because of its smooth surface, water absorption of LECA aggregate is nearly equal to or somewhat more than that of ordinary aggregate; therefore, design of light weight concrete mixture with LECA aggregate is as difficult as that of ordinary aggregate. For determining the amount of each ingredient in light weight concrete mixture (along with the amount of absorbed water in light weight aggregates, especially those with too much pores with rough and angular surface, by making different mixtures) one can use the common design methods of ordinary concrete mixture.

RESULT AND DISCUSSION

From Table 1 it is observed that for control specimen strength of the concrete increases with respect to age. For 5% replacement of cement with fly ash, fine aggregate with bottom ash, and coarse aggregate with LECA the compressive strength of concrete is the same as that of control concrete. The split tensile strength slightly decreases at early age and it attains the same strength of control concrete at 56 days.

Table 1: Mechanical properties of concrete

Percentage replacement	Dry weight of specimen (cube) in Kg/m³	Compressive strength of concrete (N/mm²)			Dry weight of specimen (cylinder) in Kg	Split tensile strength of concrete (N/mm²)		
		7 days	28 days	56 days		7 days	28 days	56 days
0	9.45	17.96	26.93	26.95	14.35	1.60	2.54	2.57
5	9.18	17.94	26.89	26.97	14.20	1.53	2.52	2.59
10	8.89	17.17	25.73	25.76	13.85	1.5	2.32	2.33
15	8.54	16.06	24.09	24.11	13.60	1.44	2.17	2.18
20	8.41	13.41	20.10	20.13	13.40	1.4	2.11	2.12
25	8.31	11.32	16.96	16.97	13.15	1.35	2.05	2.06
30	8.24	10.19	15.26	15.23	12.72	1.31	1.96	1.98
35	8.13	9.73	14.57	14.58	12.34	1.26	1.90	1.92

It is also observed that when the replacement of material increases, the compressive strength and split tensile strength decrease. The dry weight of cube and cylinder specimens decreases with respect to more replacements of materials.

Strength Analysis with respect to Age of Concrete

In Table 1 compressive strength of concrete and split tensile strength of concrete are evaluated by means of various mixing percentage applied to form cubic dry weight specimen and cylindrical dry weight specimen, respectively, with respect to different days.

For M_{20} grade concrete, the following proposition percentage mixing is taken into account for various dry weight specimens applied to cubic shape for finding compressive strength with respect to 7, 28, and 56 days such that dry weight specimen was applied to cylinder shape with respect to aforementioned days to find split tensile strength. For both strengthening analyses M_{20} grade type concrete is utilized. From Table 1 the stated results show that mixing percentage increases with decrease in specimen weight, but in strength point of view the increase in mixing percentage will certainly reduce the strength attain in both compressive strength and split tensile strength, or on other hand when the mixing proportion does not take part in this (i.e., when it is "Zero"), then the weight of the specimen is high compared to the fact that to the mixing proportion which is blended. In both cases of strength analysis extension of days will certainly the strength of projection of those analyses as clearly statedin Table 1.

Figure 3 shows the cube compressive strength analysis that takes part in three stages of consecutive days 7, 28, and 56 based on various mixing propositions The attained results show that the process done for consecutive 56-day test results shows better compressive strength on nonmixing whereas case of gradual increasing in mixing percentage will certainly reduce the compressive strength of all testing days specimens. In case of weight the increase in mixing percentage will reduce the weight.

(a)

(b)

Figure 3: Compressive strength (a) Compression test on cube, (b) Compressive strength.

Figure 4 shows the cylindrical shape split tensile strength analysis for different days. Moreover, in this split tensile strength analysis the increase in mixing percentage will certainly reduce the weight and will also reduce the strengthening factors.

(a)

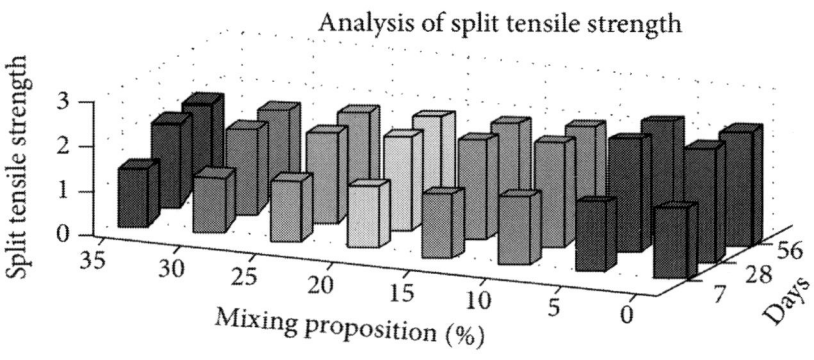

(b)

Figure 4: Split tensile strength (a) Split tensile strength on cylinder, (b) Split tensile strength.

From the aforementioned two shapes (cubic and cylinder shapes) the projected result analyses of the compressive strength and split tensile strength analyses are almost similar. Let us see the exponential behaving and its regression equation for compressive strength and split tensile strength.

Exponential Graph Based on Mixing Percentage for Compressive

Strength. Figure 5 simulates regression based exponential curve for compressive strength analysis for various mixing percentages. From Figure 5consecutive specimen test for 28 and 56 days produced almost the same values, whereas exponential equation of compressive strength in Table 2 ranges from 0 to 35 N/mm² in all four evaluation equations causing an increase in mixing percentage which will reduce all four parameters of dry weight for 7, 28, and 56 days. In the four cases other than dry weight the performance reduces, whereas in case of dry weight increase in mixing percentage will certainly reduce the weight.

Table 2: Regression equation for compressive and tensile strength

Particulars	Exponential regression for compressive strength	Exponential regression for split tensile strength
Dry weight	$Y = 9.309\,e^{-0.004x}$	$Y = 14.468\,e^{-0.0004x}$
7 days	$Y = 19.746\,e^{-0.02x}$	$Y = 1.5948\,e^{-0.007x}$
28 days	$Y = 29.61\,e^{-0.02x}$	$Y = 2.5476\,e^{-0.009x}$
56 days	$Y = 29.666\,e^{-0.02x}$	$Y = 2.5827\,e^{-0.009x}$

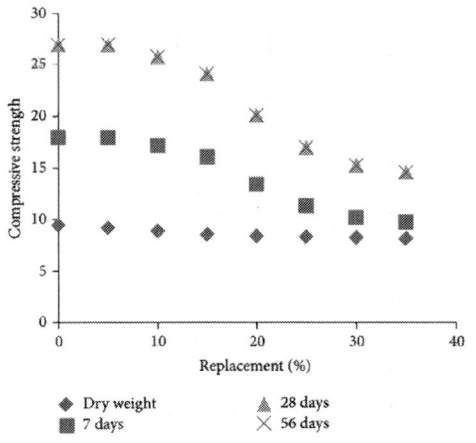

Figure 5: Exponential graph for compressive strength and dry weight.

Exponential Graph Based on Mixing Percentage for Split Tensile Strength. In Figure 6 the graph shows the exponential variation of dry weight and for various consecutive days such as 7, 28, and 56. In this dry weight having tensile strength of nearly $14.468e^{-0.0004x}$, x denotes mixing percentage; in addition to this, all other consecutive days based exponential curve get reduced and they are almost similar to each other having range of (0–15) N/mm².

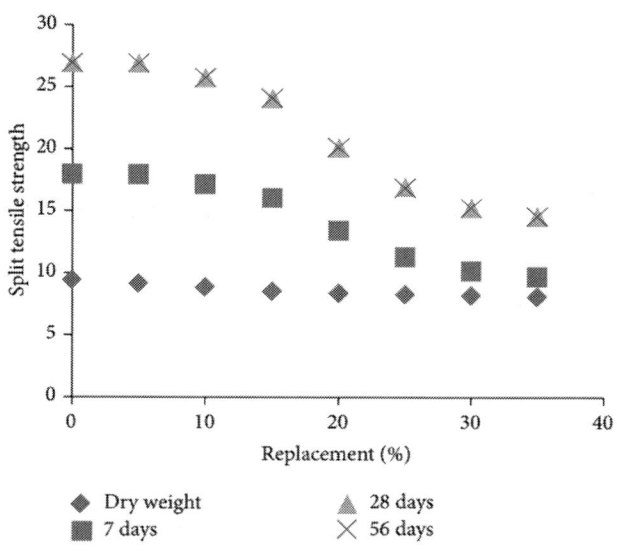

Figure 6: Exponential graph for split tensile strength and dry weight.

Table 2 includes particulars of dry weight and specimen for consecutive days such as 7, 28, and 56 days starting from dry weight in compressive strength which starts with lower regression values and keeps on increasing for 7, 28, and 56 days, whereas in the case of split tensile strength dry weight regression value is greater than the compressive strength regression value. In case of days analysis the regression values increase with the increase in the number of days in the tensile strength regression analysis model.

Flexural Strength Analysis

One measure of the tensile strength of concrete is flexural strength. It is the computation of an unreinforced concrete beam or slab to resist failure in bending (Figure 7). Designers of pavements use a theory based on flexural strength; therefore laboratory mix design based on flexural strength test may be needed. In Table 3percentages of replacement of cement with fly ash, fine aggregate with bottom ash, and coarse aggregate with light expanded clay aggregate (LECA) at the rates of 0% and 5% are employed.

Table 3: Flexural strength of beam

Specimen type	Dry weight of specimen in Kg	Flexural strength of beam (N/mm²)		
		7 days	28 days	56 days
Control	56.25	16.65	24.7	25.83
5% replacement	55.13	17.58	26.03	27.13

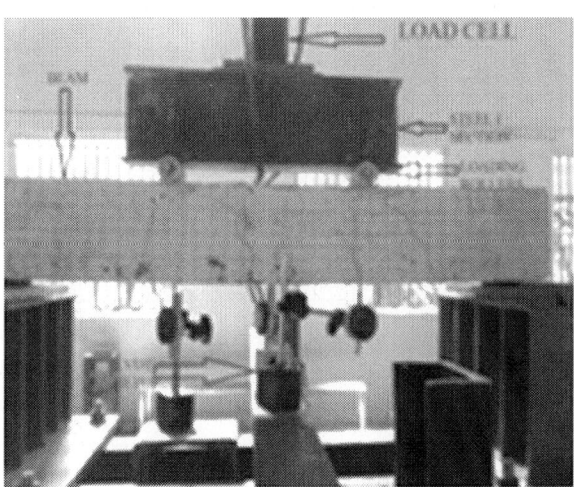

Figure 7: Flexural strength on beam.

From Table 3 the results show that the percentages of replacement of cement with fly ash, fine aggregate with bottom ash, and coarse aggregate with light expanded clay aggregate (LECA) at the rate of 5%

perform better than 0%. This dry weight of the specimen is reduced to 5% and moreover flexural strength of the beam for 7 days is 1.67% greater than 0%, and in 28 days it is 1.52% greater than 0% and in 56 days it is 1.46% greater than 0%.

In Table 4 test load is applied from 0 to 86.32 KN at various intervals and we tried to find the deflection of M_{20} in the left, middle, and right of the beam. The deflections in all levels are gradually increased when the applied load increases. The average deflection in the left of the beam is about 1.71 mm, whereas in middle deflection it is about 2.961 mm and in right side the deflection is about 1.810 mm.

Table 4: Load versus deflection of control specimen

Load (KN)	Deflection (mm) (0% replacement of fly ash, bottom ash, and LECA)		
	Left	Middle	**Right**
0	0	0	0
3.92	0.21	0.252	0.194
7.84	0.284	0.324	0.284
11.77	0.42	0.54	0.5
15.69	0.58	0.756	0.631
19.62	0.745	0.978	0.785
23.54	1.031	1.234	1.016
27.46	1.202	1.512	1.198
31.39	1.382	1.962	1.391
35.32	1.594	2.264	1.624
39.24	1.828	2.789	1.841
43.16	1.972	2.936	1.986
47.03	2.052	3.142	2.034
51.01	2.21	3.364	2.198
54.94	2.352	3.724	2.346
58.86	2.41	4.125	2.402
62.78	2.57	4.589	2.556
66.71	2.625	4.96	2.618
70.63	2.715	5.146	2.708
74.56	2.86	5.476	2.846
78.48	3.14	5.742	3.008
82.41	3.46	5.969	3.396
86.32	4.19	6.326	4.07

In Table 5 test load is applied in M_{20} from 0 to 86.32 KN at various intervals and the deflections were measured in the left, middle, and right of the beam. The deflections in all levels are gradually increased when the applied load increases. The average deflection in the left of the beam is about 1.782 mm, whereas in middle the deflection is about 2.960 mm and in right side the deflection is about 1.78 mm. From Table 5 it is proved that the deflection of 5% replacement of flexural strength is higher than 0% replacement.

Table 5: Load versus deflection of beam with optimum mix

Load (KN)	Deflection (mm)		
	(5% replacement of fly ash, bottom ash, and LECA)		
	Left	**Middle**	**Right**
0	0	0	0
3.92	0.205	0.25	0.207
7.84	0.29	0.321	0.285
11.77	0.45	0.536	0.458
15.69	0.54	0.76	0.535
19.62	0.81	1.02	0.793
23.54	1.037	1.231	1.037
27.46	1.198	1.507	1.20
31.39	1.375	1.96	1.379
35.32	1.584	2.265	1.582
39.24	1.815	2.785	1.816
43.16	2.05	2.937	2.02
47.03	2.07	3.14	2.05
51.01	2.15	3.361	2.17
54.94	2.38	3.72	2.38
58.86	2..46	4.118	2..47
62.78	2.56	4.587	2.54
66.71	2.61	4.95	2.615
70.63	2.69	5.143	2.69
74.56	2.84	5.472	2.838
78.48	3.11	5.74	3.115
82.41	3.4	5.965	3.4
86.32	4.05	6.321	4.05

In Figure 8, M_{20} grade 0% and 5% replacement of fly ash, bottom ash, and LECA are analysed to test their flexural strength. In the graph it is clearly stated that when load increases the deflection also increases for 0% and for 5% amid (23), and average deflection values are similar to both 0% and 5% but 0% they are slightly higher than 5%, whereas this graph have sum of all deflection levels in 1 unit. For example, here, the fact that the considered length of beam is said to be 1 meter for expermental investigation by appling "x" unit of load will cause the amount of deflection in both cases (0% and 5%) in respect of increase in load to certainly increase the deflection.

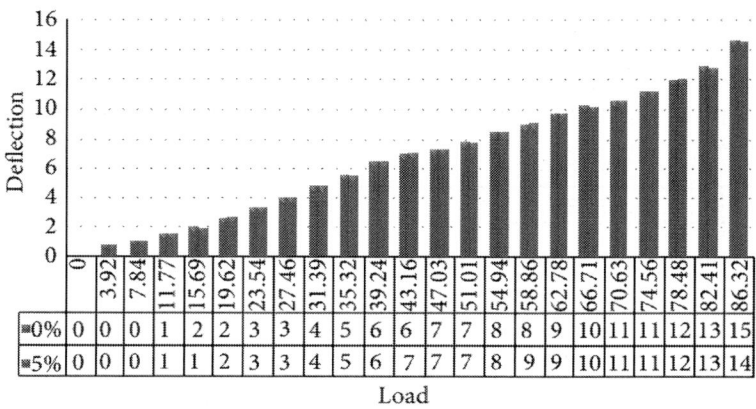

Figure 8: Load versus deflection.

CONCLUSIONS

The paper attains the highest possible strength for LECA concrete while noting the advanced technology in producing light weight concrete. The results show that 5% replacement of cement with fly ash, fine aggregate with bottom ash, and coarse aggregate with light expanded clay aggregate (LECA) was found to be good performance in compressive strength, split tensile strength, and flexural strength of beam in 56 days when compared with 28 days strength. At the same time 28 days strength also approximately equals normal conventional concrete; that is, 0% replacement and dry weight of specimen have been reduced. In future, soft computing techniques will lead with core areas us to attain better performance in short interval of time as the time is the major factor involved in this research work.

REFERENCES

1.   V. Patel and N. Shah, "A survey of high performance concrete developments in civil engineering field," Open Journal of Civil Engineering, vol. 3, no. 2, pp. 69–79, 2013.

2.   M. M. Al Bakri, H. Mohammed, H. Kamarudin, I. K. Niza, and Y. Zarina, "Review on fly ash-based geopolymer concrete without Portland Cement," Journal of Engineering and Technology Research, vol. 3, no. 1, pp. 1–4, 2011.

3.   C. Marthong and T. P. Agrawal, "Effect of fly ash additive on concrete properties," Journal of Engineering Research and Applications, vol. 2, no. 4, pp. 1986–1991, 2012.

4.   A. Charif, M. J. Shannag, and S. Dghaither, "Ductility of reinforced lightweight concrete beams and columns," Latin American Journal of Solids and Structures, vol. 11, no. 7, pp. 1251–1274, 2014.

5.   C. Meng and Z. Jin-Yang, "Studies on lightweight high-strength shotcrete," in Proceedings of the 4th International Conference on Digital Manufacturing and Automation (ICDMA '13), pp. 1231–1234, June 2013.

6.   A. A. Adegbola and M. J. Dada, "Development of mathematical equations and programs for the optimization of concrete mix designs," Journal of Science & Technology, vol. 5, no. 11, pp. 1–18, 2012.

7.   F. N. Okonta, "Frictional resistance of coal dust fouled uniformly graded aggregates," International Journal of the Physical Sciences, vol. 7, no. 23, pp. 2960–2970, 2012.

8.   H. K. Kim, J. H. Jeon, and H. K. Lee, "Workability, and mechanical, acoustic and thermal properties of lightweight aggregate concrete with a high volume of entrained air," Construction and Building Materials, vol. 29, pp. 193–200, 2012.

9.   O. Gencel, F. Koksal, C. Ozel, and W. Brostow, "Combined effects of fly ash and waste ferrochromium on properties of concrete,"

Journal of Construction and Building Materials, vol. 29, pp. 633–640, 2012.

10. M. Al Bakri, H. Kamarudin, M. Bnhussain, I. Khairul Nizar, A. R. Rafiza, and Y. Zarina, "Microstructure of different NaOH molarity of fly ashbased green polymeric cement," Journal of Engineering and Technology Research, vol. 3, no. 2, pp. 44–49, 2011.

11. A. U. Abubakar and K. S. Baharudin, "Properties of concrete using tanjung bin power plant coal bottom ash and fly ash," Journal of Sustainable Construction Engineering & Technology, vol. 3, no. 2, pp. 1–14, 2012.

12. R. J. Haynes, "Reclamation and revegetation of fly ash disposal sites—challenges and research needs,"Journal of Environmental Management, vol. 90, no. 1, pp. 43–53, 2009.

13. A. Sarkar, R. Rano, G. Udaybhanu, and A. K. Basu, "A comprehensive characterisation of fly ash from a thermal power plant in Eastern India," Fuel Processing Technology, vol. 87, no. 3, pp. 259–277, 2006.

14. T. Uygunolu, I. B. Topcu, O. Gencel, and W. Brostow, "The effect of fly ash content and types of aggregates on the properties of pre-fabricated concrete interlocking blocks (PCIBs)," Construction and Building Materials, vol. 30, pp. 180–187, 2012.

15. T. Mehmannavaz, M. Ismail, S. Radin Sumadi, M. A. Rafique Bhutta, M. Samadi, and S. M. Sajjadi, "Binary effect of fly ash and palm oil fuel ash on heat of hydration aerated concrete," The Scientific World Journal, vol. 2014, Article ID 461241, 6 pages, 2014.

16. M. R. Moini, A. Lakizadeh, and M. Mohaqeqi, "Effect of mixture temperature on slump flow prediction of conventional concretes using artificial neural networks," Australian Journal of Civil Engineering, vol. 10, no. 1, pp. 87–98, 2012.

17. I. T. Yusuf and Y. A. Jimoh, "The transfer models of compressive to tensile, flexural and elastic properties of palm kernel shell concrete," Journal of Engineering, vol. 9, pp. 195–200, 2013.

18. A. Golroo and S. L. Tighe, "Pervious concrete pavement performance modelling using the bayesian statistical technique," Journal of Transportation Engineering, vol. 138, no. 5, pp. 603–609, 2012.

19. A. Sivakumar and P. Gomathi, "Pelletized fly ash lightweight aggregate concrete: a promising material," Journal of Civil Engineering and Construction Technology, vol. 3, no. 2, pp. 42–48, 2012.

20. U. A. Abdulhameed and B. K. Salleh, "Compressive strength of high volume coal bottom ash utilization as fine aggregate in fly ash-cement blended concrete," Journal of Engineering & Technology Sciences, vol. 1, no. 4, pp. 226–239, 2013.

21. J. H. Pramana, A. A. A. Samad, A. M. A. Zaidi, and F. V. Riza, "Preliminary study on lightweight concrete under ballistic loading," European Journal of Scientific Research, vol. 44, no. 2, pp. 285–299, 2010.

22. J. J. Del Coz Díaz, F. P. Álvarez-Rabanal, O. Gencel et al., "Hygrothermal study of lightweight concrete hollow bricks: a new proposed experimental-numerical method," Journal of Energy and Buildings, vol. 70, pp. 194–206, 2014.

23. E. Güneyisi, M. Geso□lu, and S. Ipek, "Effect of steel fiber addition and aspect ratio on bond strength of cold-bonded fly ash lightweight aggregate concretes," Construction and Building Materials, vol. 47, pp. 358–365, 2013.

24. P. Posi, C. Teerachanwit, C. Tanutong et al., "Lightweight geopolymer concrete containing aggregate from recycle lightweight block," Materials and Design, vol. 52, pp. 580–586, 2013.

25. A. U. Abubakar and K. S. Baharudin, "Potential use of malaysian thermal power plants coal bottom ash in construction," Journal of Sustainable Construction Engineering & Technology, vol. 3, no. 2, pp. 1–13, 2012.

Chapter 2

Separation Strategies for Processing of Dilute Liquid Streams

Sujata Mandal[1] and Bhaskar D. Kulkarni[2]

[1]Expertise Center for Eco-Testing Laboratory, Central Leather Research Institute, Adyar, Tamilnadu, Chennai 600020, India

[2]Chemical Engineering and Process Development Division, National Chemical Laboratory, Dr. Homi Bhabha Road, Maharashtra, Pune 411008, India

ABSTRACT

Processing of dilute liquid streams in the industries like food, agro-, biotechnology, pharmaceuticals, environment, and so forth needs special strategy for the separation and purification of the desired product and for environment friendly disposal of the waste stream. The separation strategy adopted to achieve the goal is extremely important from economic as well as from environmental point of view. In the present paper we have reviewed the various aspects of some selected universal separation strategies such as adsorption, membrane separation, electrophoresis, chromatographic separation, and electroosmosis that are exercised for processing of dilute liquid streams.

INTRODUCTION

Processing of dilute liquid streams occurs in number of unit processes and unit operations in chemical industries. Whereas the separation strategies adopted by these industries sometimes follow the conventional processes such as distillation, crystallization, drying, and so forth, due to the familiarity of operation since several decades. In industry, handling of dilute liquid streams needs special strategies for the separation and purification of the desired product and for environmental friendly disposal of the waste stream. The separation strategy adopted is extremely important from economic as well as from environmental point of view. The suitability of a separation technique depends on number of factors that includes:

- improved selectivity,
- improved energy efficiency,
- development of new process configurations and integration,
- economic viability,
- environmental safety and compatibility, and
- sustainability (recycle and reuse).

Numerous new/modified techniques have been exercised by engineers and scientists to improve the efficiency and reduce the cost of the traditional separation techniques. Separation processes of commercial interests can be categorized according to the phases involved as

- solid-solid separation (screening, classification, floatation, flocculation, and field-based),
- solid-liquid separation (thickening, centrifugation, filtration, drying, and crystallization),
- solid-gas separation (cyclone, filters, adsorption, etc.),

- liquid-liquid separation (distillation, extraction, membranes, and adsorption),
- liquid-gas separation (absorption, stripping, and pervaporation),
- gas-gas separation (membranes), and
- solid-liquid-gas separation.

Each of these categories has several alternative ways to bring about separation. However, it is not possible to look at all such possible strategies exhaustively within a single frame. In the present paper, we wish to look into the various aspects of the different universal separation strategies used for processing of dilute liquid streams, such as, adsorption, membrane separation, electrophoresis, chromatographic separation, and electroosmosis.

SEPARATION METHODOLOGIES

Adsorption

Adsorption is a thermodynamically spontaneous surface phenomenon. When a multicomponent liquid mixture is contacted with a solid surface (adsorbent), certain component of the mixture (adsorbate) gets concentrated at the surface of the solid due to difference in the intermolecular forces of attraction between the components of the liquid mixture and the solid. This formation of adsorbed phase on the surface of the solid adsorbent having a composition different from that of the bulk liquid phase forms the basis of separation by adsorption technology. The process of adsorption is an exothermic phenomenon while it's reverse, that is, desorption, is an endothermic phenomenon. Both adsorption and desorption forms equally essential steps in a practical adsorption process where the adsorbent is repeatedly used for carrying out the separation.

Adsorption technology is considered as an option for separation when high degree of purity is required. Adsorptive separation could provide an alternative to

- systems for energy intensive cryogenic distillation and liquefaction,
- conventional distillation systems,

- systems require minimum use of inprocess air and water, and
- complex separation involving high boiling or thermally unstable compounds.

Development of adsorption technology primarily depends on the development and utilization of wide variety of micro- and mesoporous adsorbents with varying pore structure and surface properties, which are responsible for the selective adsorption of specific component from liquid mixtures. The usefulness of an adsorbent in a separation and purification process is a function of its composition, its pore structure and surface properties, the presence and type of functional groups at the surface, its degree of polarity, and its hydrophilic/hydrophobic characteristic. The adsorbent should not be vulnerable to fouling or degradation by the components of the feed stream. The commonly used adsorbents include aluminosilicates (zeolites and molecular sieves), activated carbon, natural and synthetic clays, activated and impregnated aluminas, silica gels, ion-exchange resins, and biopolymeric adsorbents. Each of these groups of adsorbents contains number of subgroups of materials. Development of new adsorbent materials and physicochemical modifications of the existing adsorbents is being continuously studied by the researchers.

The adsorption efficiency of a specific adsorbent-adsorbate system depends on a number of factors:

- the physicochemical nature of the adsorbent—chemical composition, surface properties, and pore structure,
- the nature of the adsorbate such as its pKa, functional groups present, polarity, molecular weight, and size, and
- the solution conditions such as solution pH, ionic strength, and the adsorbate concentration.

Modes of Operation of Adsorption Process

The adsorptive separation process can be performed through various modes.

Batch Adsorption Process

The batch process is the intuitive way of studying the efficiency of an adsorbent for adsorptive separation of a specific molecule or ion. Here, a fixed amount of the adsorbent is contacted with a fixed volume of the adsorbate solution of known concentration at constant temperature. The change in concentration of the adsorbate after a fixed time interval gives the adsorption capacity of the adsorbent for the selected adsorbate. This is the first step to design an adsorptive separation process.

Continuous Flow Adsorption Process

In this process, the feed stream is percolated through a fixed-bed column packed with the adsorbent in a continuous flow mode. On near saturation of the adsorbent, as evidenced by breakthrough of adsorbate, the column is removed and regenerated. There are two basic modes for operation of bed or column type adsorbers relating to exhaustion and regeneration of adsorbents, namely, fixed beds and pulsed beds. For high concentration systems typical of certain industrial wastes, pulsed-bed systems are often effective and efficient. In continuous flow process, the reactor can be designed to operate in either upflow or downflow direction. A fixed-bed adsorber can be operated in the upflow mode to minimize pressure drop, channeling and fouling of the adsorbent. In addition, upflow design and operation allows smaller particle sizes of adsorbent to be employed to increase adsorption rate and hence decrease adsorber size.

Adsorption Chromatography

In this process, the feed stream is introduced as a pulse in a purge stream, similar to liquid phase chromatography. This method has been applied for small-scale production of fine chemicals. There are limitations in large-scale separation by adsorption chromatography.

Applications of Adsorption in Separation and Purification

The field of adsorption technology has grown extensively over the past years, and it will continue to grow in the prospective future because of its vast applicability. The endless choice of adsorbent materials and their use in the design of innovative separation processes provide bright future for this technology as a separation tool. Adsorption technology has found numerous applications for separation and purification of dilute liquid mixtures and for product recovery in the chemical, petrochemical, and biochemical industries, in biotechnology and biomedical applications and for water treatment (ground/surface water and effluent). The technology has been used for olefin-paraffin separation, fructose and glucose separation, isomer separation (xylene, cresol, and cymene), for breaking azeotropes, recovery of antibiotics, removal of microorganisms, purification and recovery of biomolecules (proteins, vitamins, and enzymes), trace impurity removal, for municipal and industrial waste treatment, ground and surface water treatment, to name but a few. Development of new adsorbents and advanced adsorption methods specialized in compositions, structures, functions, and characteristics made this technology suitable to meet the industrial needs of modern era.

An overview of the separation and purification of different valuable products from dilute solutions and removal of various contaminants from dilute process streams by adsorption technique has been presented.

Industrial Applications

For large-scale industrial separation from dilute liquid streams, adsorption technique has been found to be more sustainable alternative in terms of low energy costs and process economics. One component can selectively be adsorbed in a bed of adsorbent particles, while the other passes through. The adsorbed component can be recovered afterwards by thermal swing adsorption (TSA) or pressure swing adsorption (PSA) depending on the nature of the adsorbent and adsorbate. In this way, the separation can be achieved at ambient temperature. Due to size-selective adsorption characteristics, different forms of zeolite and molecular sieves have gained considerable interest towards their industrial applications [1] for the separation of light olefin/paraffin mixtures [2], glucose-fructose mixtures [3], isomer separation [4, 5], breaking azeotropes [6], separation of aromatics for the purpose of purification of chemicals [7], and so forth. In addition to various forms of zeolite, polymeric adsorbents and advanced materials containing metal-organic framework (MOF) have been developed for adsorptive separation of aliphatic and aromatic isomers [8].

Continuous separations of olefin/paraffin mixtures are of high demand in chemical industries. Herden and his coresearchers have extensively studied the separation of olefin/paraffin mixtures using different modified forms of X- and Y-type zeolites [2, 9, 10]. The studies revealed that X-zeolites (Si/Al ratio 1.3) are more appropriate for olefin/paraffin separation. Recent development in this area has shown that adsorbents containing transition metal salts dispersed on high surface area substrates like zeolites, alumina, ion-exchange resins, silica, clays, and activated carbon are more efficient for enhanced separation of olefin/paraffin mixtures in bulk [11–13]. This type of adsorbent comprising of transition metal salts supported on high surface area substrates (zeolite, activated alumina, and silica) are also found useful for desulfurization of commercial diesel fuel [14, 15] and adsorptive removal of aromatics [16] in industries.

Biotechnology and Biomedical Applications

Separation and purification of biomolecules like, proteins, vitamins, enzymes, and antibiotics from their dilute solutions is an important task in bimedical and biotechnological applications. In this field of application, the primary focus is given on selectivity of the adsorbent for the specific molecule as well as easy recovery of the molecule from the adsorbent. Studies reporting the use of a wide variety of adsorbents including inorganic mesoporous silica, aluminosilicates, carbon nanotubes, natural and synthetic polymeric adsorbents, polymeric resins, and so forth are reviewed. Different modes of the adsorption process like batch process and continuous flow process are opted to achieve the separation and purification of the desired product.

Being a low-energy consuming and high-yield process, adsorptive separation and recovery of biomolecules, namely, antibiotics and vitamins, from fermentation broth are of great industrial importance. Adsorbents such as nonionic polymeric resins [17], ion-exchange resins [18, 19], activated carbon, molecular sieves, and so forth have been reported to be effective for isolation and recovery of these biomolecules. Separation of -lectum antibiotics, such as amoxicillin, ampicillin, cephalexin and cefadroxil, trimethoprim, from bioreactors are achieved by adsorption on polymeric resins, activated carbon and biopolymeric adsorbents [20, 21].

DNA separation by adsorption on pure/modified silica and its composites is an important method used in novel technologies that uses microchannels [22]. The principle behind this type of separation relies on DNA molecules binding to silica surfaces in the presence of certain salts and under certain pH conditions [23]. The other adsorbents used for separation of DNA molecules by adsorption are natural/modified clays [24], aluminosilicates, oxidic minerals like goethite [24] and magnetic nanoparticles, and carbon nanotubes [25]. Separation and purification of proteins is usually carried out by adsorption onto different adsorbents materials packed in a chromatographic column. Adsorbents such as polymeric/composite ion exchangers [26], magnetic nanoparticles [27], cryogel [28], polypyrrole-based adsorbents [29], and mesoporous SBA-15 [30] have been developed and used for separation of proteins. Inorganic and organic-polymeric membranes are also studied by researchers for adsorptive separation of protein molecules [31, 32]. A new concept for the purification of protein in micellar aqueous two-phase system using magnetic adsorbent has been reported by Becker et al. [33].

Water Treatment

Adsorption technology is extensively exploited by several researchers for water treatment and its purification [34]. Though a wide variety of adsorbents have been studied for water treatment by adsorption, in practice, a few of them have been practically utilized for the purpose. Adsorbent cost is the key point while selecting the adsorbent for this specific application. The selection of adsorbent is done primarily on the basis of the quality of the required water (drinking/industry/ environmentally safe wastewater) and the adsorbent cost. Activated carbon (AcC) obtained from different sources [35, 36], activated and impregnated alumina [37, 38], natural-modified clays [39] and minerals [40, 41], synthetic clays [42, 43], synthetic and modified zeolites [44, 45], polymers, and ion-exchange resins [46, 47] are among the most exploited adsorbents for water treatment. Being low cost and capacity to adsorb both organic and inorganic contaminants, activated carbon adsorbents and their chemically modified forms are brilliant for ground/surface water as well as wastewater treatment [48, 49]. Activated carbon (AcC) is usually available in four different forms, powder (PAC), granular (GAC), fibrous (ACF), and clothe (ACC). All these four forms of activated carbon obtained from various sources have been investigated for drinking water as well as wastewater purification. Adsorption capacities of activated carbon for various contaminants in aqueous medium (as reported in literature) have been listed in Table 1. An extensive survey of literature on the removal of inorganic contaminants from water reveals that activated carbon adsorbents are not very effective in removing inorganic contaminates including heavy metals from water/wastewater. In this context, a large number of low-cost materials that include agricultural wastes, natural bio-polymers (chitosan, alginic acid, and cellulose), industrial wastes/biproducts, natural clays, soils and minerals, and their chemically modified forms are studied by several researchers and found to show excellent adsorption property for heavy metal removal as well as for the removal of colorants/dyes from water/wastewater [65–68]. Natural soils have been successfully used by researchers for the removal of inorganic contaminants like, fluoride [69], arsenic [70, 71], iodate [72], phosphates [73], and so forth from surface/ground water. Unfortunately, very little information is available in the literature regarding the recovery of pollutants from these low-cost adsorbent materials, the regeneration of spent sorbents, and the stability and reproducibility of the sorbents. Hence, further research is required for real-life applicability of these low-cost adsorbents.

In addition to the traditional adsorbents, researches are going on for the development of new effective adsorbents that can be regenerated and reused for cost-effective separation process. Many advanced adsorbent materials such as carbon-natural zeolite composite [74], aluminium-loaded shirasu zeolite [75], carbon nanotubes [76], alumina-supported carbon nanotubes [77], zeolite-portland cement mixture [78], zeolite-vermiculite composite [79], cellulose-supported synthetic clay [80], carbon-fly ash composite, thiol-functionalized silica coated activated alumina [81], Ce-Ti oxide [82], and so forth with improved adsorption characteristics are being developed and tested for application in water treatment.

Over the last decade, large progress has been made toward the process development of this technology, but, unfortunately, the development in the field of adsorbent materials is not significant. Successful implementation of the technology for separation of industrial processed streams requires reasonably good understanding of the fundamentals of adsorption and adsorption processes, that is, forces responsible for adsorption, adsorption equilibrium, its temperature dependence, heat of adsorption, process cost, and so forth. The technical challenges involved for further growth in this area is the cost-effective quality separation, which require more efficient way to regenerate and reuse adsorbents, to improve our understanding in tailoring adsorbents for complex system and to improve predictive models for mass transfer, adsorption, equilibrium, and other physical data.

Table 1: Adsorption characteristics of activated carbon derived from different sources, reported in the literature, for the removal of various contaminants from water at 25–30°C

Carbon type and source	Activation	S.A.1 (m2/g)	Adsorbate	Initial conc. (mg/L)	Ads. capacity (mg/g)	Model used	Ref.
GAC-coal	Untreated	689	Aniline	250–3000	242.92	Langmuir	[50]
GAC-coal	Untreated	689	Pyridine	250–3000	157.01	Langmuir	[50]
GAC-coal	Untreated	689	Phenol	250–3000	218.97	Langmuir	[50]
GAC-coal	Untreated	689	Benzene	250–3000	271.51	Langmuir	[50]
AcC-waste tire	HF, heating at 850°C, chlorination	1060	Toluenea	500	421.8	—	[51]
AcC-pine bark	Heating at 672°C under N2flow	332	Phenol	99.75	28.20–40.42	Langmuir	[52]
ACC	Conductivity water at 60°C under N2 flow	2500	Pesticide (ametryn)	103.84	354.61	Langmuir	[53]
ACC	-Do-	2500	Pesticide (aldicarb)	103.96	421.58	Langmuir	[53]
ACC	-Do-	2500	Pesticide (dinoseb)	22.38	301.84	Langmuir	[53]
ACC	-Do-	2500	Pesticide (diuron)	32.27	213.06	Langmuir	[53]
ACC	-Do-	2500	Pesticide (bentazon)	20.90	151	Langmuir	[54]
ACC	-Do-	2500	Pesticide (propanil)	23.52	114	Langmuir	[54]
PAC-Bokbunja waste seeds	Untreated		Dye-procion red MX-5B	10–60	29.37	Langmuir	[55]
PAC-Bokbunja waste seeds	n-hexane		-Do-	10–60	30.65	Langmuir	[55]

AcC-sunflower seed hulls	H2SO4	135.2	Dye-acid blue 15	—	125	Langmuir	[56]
AcC-pine-fruit shell	K2CO3	1035	Brilliant green dye	600	216.1	Langmuir	[57]
AcC-pine-fruit shell	Heating at 600°C	1425	-Do-	600	284.3	Langmuir	[57]
AcC-water hyacinth	H2SO4 followed by heating at 300°C	—	Fluoride	2–25	6.06	—	[58]
AcC-water hyacinth	H2SO4 followed by heating at 600°C	—	-Do-	2–25	4.66	—	[58]
AcC-commercial (coconut)	Untreated	1050	NH3	340.1	5.26	—	[59]
AcC composite	Sodium alginate + chitosan + activated carbon	—	Hg2+	—	576	—	[60]
AcC-coconut shell	Untreated	1280	Cr(VI)	120	107.1	—	[61]
AcC wood	Untreated	1700	Cr(VI)	120	87.6	—	[61]
AcC coal	Untreated	1120	Cr(VI)	120	101.9	—	[61]
GAC	Water wash	648	Cu(II)	—	6.14	Langmuir	[62]
GAC	Citric acid	431	Cu(II)	—	14.92	Langmuir	[62]
GAC	Citric acid followed by NaOH	448	Cu(II)	—	11.85	Langmuir	[62]
AcC-rubber leaf	HCl	0.57	Cu(II)	5–50	8.39	Langmuir	[63]
AcC-saw dust	Untreated	686.3	Pb(II)	51.8–414.4	84.95	Langmuir	[64]

[a]Adsorption experiments were performed at 22°C.

[1]Surface area.

Membrane Separation

During the past decades, membrane separation processes have been developed and optimized for large-scale industrial applications for separation and purification of desired products present in dilute solutions. In this separation technique, the liquid stream is passed through the membrane module when the solvent molecules pass through the membrane leaving desired product/molecule on the other side. This technology is generally used in bulk rather than precise separation.

The most significant advantage of membrane-based separation systems over other existing separation processes is the consumption of less energy. In addition to energy saving, membrane systems are compact and modular, which facilitate the system an easy retrofit to existing industrial processes. The technology is extensively used in food and beverage industry [83, 84], pharmaceutical industries, biotechnology [85, 86], in environmental, in the treatment of effluent and process streams [87], and in energy applications. The applications of different membrane-based separation processes used in various industries are discussed as follows.

Microfiltration and Ultrafiltration (MF and UF)

This is a process for separating materials of colloidal size and larger from true solutions using polymeric membranes. It is used for purification of aqueous streams, concentration, purification, and recovery of valuable products. This is the most widely used among all other membrane processes.

Microfiltration/ultrafiltration (MF/UF) technique is mostly used for aqueous fluid treatments. The application mechanism comprises of purification by the rejection of impurities (e.g., purification of water), by retention and concentration of valuables (purification of biomolecules), and by permeation and purification of valuables (purification of fermentation fluid).

The two most important characteristics for a successful MF/UF process are pore size and economic efficiency of the membrane. For industrial application, the pore size of the membrane must be appropriate for the purpose of separation, and the economic efficiency must be sufficient to allow for the variations in pore sizes needed.

The MF/UF process has been largely used in water treatment for treating natural water, drinking water [88, 89], wastewater, reservoir water [90], and oily water [91]. The technique is also exploited for the treatment of algae-rich water [92], the removal of virus from water and wastewater [93], the removal of colloids and natural organic matter from surface water, and so forth. Several efficient water-treatment technologies that include MF/UF process along with other processes like precipitation/coagulation [94], electroperoxidation [95], ozonation [96], photooxidation [97], and so forth have also been developed. The

MF/UF process is also a widely used method for pretreatment of water before reverse osmosis or nanofiltration. Tsai et al. [98] have reported electromicrofiltration method for the removal of natural organic matter and inorganic particles from natural surface water. In this method, the microfiltration is performed under an applied electric field when an enhanced removal of natural organic matter and inorganic particulates was obtained because of electrophoretic and electroosmotic effects.

In food processing industry, MF/UF separation technique is used for processing dairy products [99,100], honey [101], fruit/vegetable juice [102, 103], clarification of wine, and decolourization of sugarcane solution [104].

This MF/UF technique is also used for recovery of biomolecules like proteins, enzymes, vitamins, and carbohydrates from milk products [105], vegetable oils [106], fishery products [107], poultry processing wastewater [108], fermentation products [109], and peels of citrus fruits [110].

However, the main problem of MF/UF process is the flux decrease caused by concentration polarization and membrane fouling. Several processes have been tried by researchers for preventing/reducing the membrane fouling.

Reverse Osmosis (RO)

This is a liquid/liquid separation process that uses a dense semipermeable membrane, highly permeable to water and highly impermeable to microorganisms, colloids, dissolved salts, and organics. This is the first membrane-based separation process to be widely commercialized. This technique is largely used for water treatment [111, 112], namely, production of demineralized or potable drinking water, desalination of seawater and brackish water, and pure boiler water makeup in industrial fields, and in food processing industries, and wastewater treatment and reuse.

The largest single application area of reverse osmosis is desalination of seawater and brackish waters [113, 114]. Cellulose acetate membranes and thin-film composite membranes made from aromatic polyamides [115, 116] are widely used for this purpose. Several pretreatment methodologies like flocculation/precipitation, MF/UF [117], nanofiltration [118], adsorption, electrocoagulation [119], and so forth for seawater reverse osmosis are tried to increase the membrane life by reducing membrane fouling. RO separation of seawater and brackish water containing high total dissolved solid (TDS) is considered to be less energy consuming than electrodialysis. It is also used for the removal of boron from water [120].

In addition to the desalination of water, RO technique is also being used for the production of drinking water, water treatment for boilers, feed water treatment for industrial use, the removal of organic and inorganic contaminants from water, wastewater treatment and reuse [121–124].

Other applications of reverse osmosis separation technique include purification of lactic acid from fermentation broth [125], separation of organic/inorganic compounds from their aqueous solutions [126, 127], separation of organics from multicomponent mixtures [128], boric acid recovery [129], and so forth.

Electrodialysis (ED)

This is an electrochemical separation process in which ionic species are separated from an aqueous solution or from other uncharged components using electrically charged membranes under the driving force of electrical potential difference. Though the separation technique was first developed for the desalination of brackish water as can be seen from the recent works of Sadrzadeh and Mohammadi [130, 131] and many other research groups but at present, the most important industrial application of electrodialysis is in the production of potable water. This technique is widely used for the removal of dissolved metal ions and solids from industrial processed/waste streams and concentration of metal ions from their solutions [132, 133]. The removal of inorganic contaminants like fluoride (F^-), nitrate (NO_3^-) [134], and boron [$B(OH)_4^-$] from dilute aqueous streams are also performed by electrodialysis method [135] through different ion-exchange membranes [136]. It can also be utilized for separation and concentration of salts [137], acids (mineral and organic) [138–140] and bases from aqueous solutions, separation of monovalent ions from multiple charged components [141], and separation of ionic compounds from uncharged molecules [142]. Due to the diversity and practicability of the technique, it can be a versatile tool to meet specific needs from chemical [143], biochemical [144], biotechnological [145], food [146], and pharmaceutical industries [147, 148]. Electrodialysis-based separation techniques have been reviewed by Xu and Huang [149]. In recent days, slightly modified form of electrodialysis with ultrafiltration membrane has been developed to separate valuable biolmolecules [150] on the basis of their electrical charge and size or molecular weight. In this process, a conventional electrodialysis cell is used, in which some ion exchange membranes are replaced with ultrafiltration membranes [151], compounds of higher molecular weight than the membrane cutoff can be separated so as to extend the field of application of electrodialysis to biological charged molecules.

Pervaporation

Pervaporation is a membrane-based separation technique used for separation and concentration of liquid mixtures, especially of aqueous-organic azeotropes [152]. In this technique, the liquid mixture to be separated get contacted on one side of the membrane, and the permeate is removed as a vapor from the other side. The membrane acts as a selective barrier between the two phases, the liquid feed solution and the vapor phase permeate. Transport through the membrane is induced by the difference in partial pressure of the components on the two sides (the feed solution and the permeate vapor phase) of the membrane. The mass transport across the membranes involves three successive steps.

- Upstream partitioning of the feed components between the flowing liquid feed and the upstream surface layer of the membrane.
- Diffusion of the components through the membrane.
- Desorption of the components at the permeate side of the membranes.

The steady-state mass transport system depends on the parameters, like, upstream pressure, downstream pressure, temperature, and film thickness. When the downstream pressure is low, the flux is inversely proportional to the film thickness. Though the first industrial application of pervaporation was for the dehydration of alcohol/water azeotropic mixtures, presently the technology finds wide industrials applications for separation of and recovery from various liquid mixtures. For being economical, safe, and eco-friendly process, pervaporation separation is considered to be a promising alternative to conventional energy intensive separation technologies, like extractive or azeotropic distillation in liquid mixtures. Pervaporation can be considered the so-called "clean technology," especially for the separation of volatile organic compounds [153, 154]. The technology is highly used for the separation of organics like aliphatic and aromatic alcohols, acids, benzene, toluene, tetrahydrofuran, ethylene glycol, and so forth from dilute aqueous solutions [155, 156]. It is also extensively used for the separation of organic/organic mixtures [157], organic azeotropic mixtures [157–159], and separation of isomers [160, 161]. The technique has been studied for the recovery of natural aroma compounds in food industry [162].

Natural biopolymers like chitosan and sodium alginate have attracted considerable attention during the last decade as membrane material for pervaporation separation [163–165]. Due to hydrophilic nature and ability to modify/tune their structures to achieve the desired separation, sodium alginate has fascinated wide group of researchers. Sodium alginate is the widely studied membrane in pervaporation dehydration of industrial solvents like ethanol, isopropanol, acetic acid, tetrahydrofuran, and so forth. The performances of sodium alginate-based membranes are reported to be outstanding in dehydration of organics from aqueous-organic mixtures. Use of sodium alginate membrane and its various modified forms in pervaporation dehydration studies has been increasing over the years. Figure 1 shows the trend of published literature on the use of sodium alginate-based membranes for pervaporation separation of aqueous-organic mixtures for past few years.

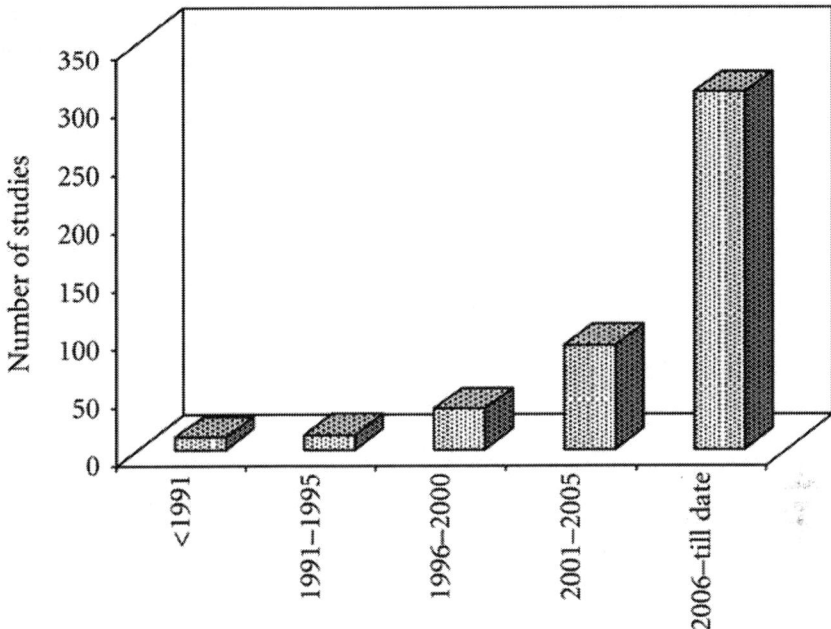

Figure 1: Number of studies reported in literature on the use of sodium alginate-based membranes for pervaporation separation of aqueous-organic mixtures (sources used: ScienceDirect online search, online ACS journals, online library of wiley publication and online journals of Taylor and Francis Publication).

Aminabhavi and his research group have extensively studied the application of sodium alginate and its modified membranes in pervaporation separation of aqueous-organic mixtures. Different types of blend, grafted, and mixed matrix sodium alginate membranes are prepared and tested for their pervaporation performances by the research group of Aminabhavi. A comparison in the pervaporation performance of sodium alginate-based membranes with that of other polymeric membranes for the separation of water-isopropanol mixture at 30°C has been listed in Table 2. The data presented are based on the reported works by Aminabhavi and his research group.

Table 2: Pervaporation performance of different types of membranes for the separation of water-isopropanol mixture with 10 wt. % feed water at 30°C

Membrane type	Selectivity	Ref.
PVA	77	[166]
PVA/PMMA blend membrane (5% PMMA)	400	[166]
PVA-PMA blend membrane (4% PMA)	2342	[167]
Pristine/blend/grafted sodium alginate membranes		
Sodium alginate (pristine)	650	[165]
Poly(acrylamide) grafted sodium alginate	100	[165]
Sodium alginate/PVA blend	580	[165]
Sodium alginate/pAAm grafted GG blend	890	[165]
Sodium alginate + 5 wt.% PVA + 10 wt.% PEG	3600	[165]
Mixed matrix membranes of sodium alginate		
MCM-41 (10 wt.%) filled sodium alginate	30,000	[165]
SBA-15 (10 wt.%) filled sodium alginate	∞	[165]
Fe-SBA-15 (10 wt.%) filled sodium alginate	∞	[165]
Na+MMT (10 wt.%) filled sodium alginate	∞	[165]
AlPO4-5 (20 wt.%) filled sodium alginate	69,000	[165]
Al-MCM-41 (20 wt.%) filled sodium alginate	∞	[165]

PVA: poly (vinyl alcohol); PMMA: poly(methyl methacrylate); PMA: phosphomolybdic acid; NaAlg: sodium alginate; pAAm: poly(acrylamide); GG: guar gum; PEG: poly(ethylene glycol); Na+MMT: sodium montmorillonite; $AlPO_4$-5: aluminophosphate.

Table 2 shows that even though isopropanol separation has yielded poor selectivity with pure sodium alginate membrane, but large improvements were observed by incorporating sodium alginate with fillers like MCM-41, SBA-15, Na+MMT, $AlPO_4$-5, and so forth. However, the use of sodium alginate membrane for separation of organic-organic mixture by pervaporation is rare.

Although pervaporation through polymeric membranes is the most exploited one, several inorganic polymeric composite membranes with higher thermal and mechanical stability have also been developed by the researchers to make the transport process faster in pervaporation

[168]. Apart from composite membranes, inorganic (zeolites and silica-based ceramic) membranes [169] also have been attracting attention of researchers in the recent years because of their tailored selectivity, high flux and low energy consumption in addition to provide a scope for combined reaction-separation systems.

Different hybrid pervaporation techniques such as pervaporation adsorption [170], distillation pervaporation [171], and so forth are also practiced for improved separation and enhanced recovery.

Simultaneous reaction separation by combination of pervaporation with chemical reactors has become an interesting alternative to the conventional process. In a pervaporation reactor, separation of product from the reaction system enhances the conversion and hence the products yield [172]. The removal of product during reaction in a pervaporation reactor suppresses the formation of byproduct. Pervaporation membrane reactor is one such reaction-separation system implemented for esterification reactions [172], production of butanol, extraction of aroma components from fermentation reactors, resin production [173], wastewater treatment, and so forth [174]. A new concept of hybrid pervaporation membrane reactor system described by Park and Tsotsis [175], which integrates pervaporation reactor system with adsorption in the permeate side. Park and Tsotsis have validated the hybrid pervaporation membrane reactor system by performing esterification of acetic acid by ethanol.

The major constraints for practical application of the membrane technology for processing of dilute liquid streams are the need for membrane that can work efficiently in real-world applications in complex matrix of the processed streams and high scaleup costs for large-scale applications.

The current needs for further development of the membrane technology include the following.

- Development of low-cost membranes with high surface area per unit volume.
- Development of high temperature membranes (ceramic/metal), nanocomposites, targeted to specific commercial applications.
- Increasing membrane life with antifouling and antiflux schemes.
- Regeneration/low-cost maintenance of the membranes.

- Integration of the membrane technology with other separation technologies.
- Improved design and information tools for predicting membrane performance.
- Process scaleup.

Electrophoresis

Electrophoretic separation may be the major technique for molecular separation in today's cell biology laboratory for analytical as well as for preparative purposes. This is an inexpensive and powerful technique for separation in molecular level. Electrophoretic separation technique is based on the differential migration of electrically charged particles in an electric field. Hence, this technique is applicable only to ionic or ionogenic materials. The mediums used in biochemical applications are usually aqueous solutions, suspensions, or gels.

The key mechanism in the theory of electrophoresis is the electrical double layer, which is formed by the fixed charges of the macromolecules (or colloid) with the relatively mobile counter ions of the surrounding fluid. The thickness of the double layer is normally given by the inverse of the Debye-Huckel constant

$$\frac{1}{\kappa} = \left(\frac{\varepsilon k T}{8\pi e^2 n_o z^2} \right)^{1/2},$$

(1)

Where e is the electronic charge, k is the Boltzmann constant, n_o is the bulk concentration of each ionic species, z is the valence of the electrolyte, and is the dielectric constant.

The total charge of the electrical double layer is zero; however, the spatial distribution of charges is not random and hence gives rise to the electric potential. The mobile counter ions are bound sufficiently tightly to the macromolecules so that, during electrophoresis, the mobile counter ions moved together with the macromolecules.

Electrophoresis can be one dimensional (i.e., one plane of separation) or two dimensional. One-dimensional electrophoresis is used for most routine protein and nucleic acid separations. Two-dimensional separation of proteins is used for finger printing. Alternatively, the electrophoretic technique may be of the following types: moving boundary electrophoresis (MBE), zone electrophoresis (ZE), disc electrophoresis, isoelectric focusing (IEF), sodium dodecyl sulphate/ polyacrylamide gel electrophoresis (SDS/PAGE), isotacophoresis (ITP), DNA sequencing, immobilized pH gradients (IPG), pulsed-field gel electrophoresis, and capillary zone electrophoresis. The latest addition in the list is chromatophoresis, which is direct coupling of HPLC with SDS/PAGE and thus provide a new type of 2-D map. Moving boundary electrophoresis, capillary electrophoresis, and zone electrophoresis are the separation techniques primarily used in analytical methods. While most of the electrophoresis techniques have been used for analytical purposes, zone electrophoresis and isoelectric focusing are the two variations, which are also useful for preparative separations of protein mixtures.

Zone electrophoresis is characterized by the complete separation of charged solutes into separate zones. Many of the earlier separations of this type were carried out using filter paper in a Durum cell. Capillary electrophoretic technique has been used for the separation of DNA, proteins, in biotechnology and cell-biology applications [176, 177]. The technique is also used for chiral separation in pharmaceutical industry [178]. Prior to dramatic improvements in chromatographic techniques, continuous electrophoresis was popular for purifying proteins such as enzymes and amino acids [179, 180]. Electrophoresis is still invaluable on an analytical scale, but large scale electrophoretic separations were not established because the technique does not translate well to large sizes, mainly because of the difficulties to remove the generated heat during the separation and also because the process is very slow. During the recent decade, lot of progress has been made in electrophoretic separation technique.

Chromatographic Techniques

Different forms of chromatographic separation techniques are widely used in separation and purification of industrial process streams, such as for purification of groundwater, separation of chiral compounds in pharmaceutical industries, and separation of biomolecules in bio-chemical/biomedical applications. The chromatographic techniques are broadly classified into two categories depending on the physical state of the mobile phase, gas chromatography (GC), and liquid chromatography (LC). As the present paper deals with separation in solution; hence, the discussion is restricted within liquid chromatography.

Liquid Chromatography (LC)

This is the chromatographic technique in which the mobile phase is a liquid. Depending on the separation mechanism, the liquid chromatography can be classified into the following subcategories.

High-Performance Liquid Chromatography (HPLC)

This is a highly improved form of column chromatography. In this technique, the liquid mobile phase along with the sample to be separated is moved through the stationary phase (packed column) under high pressure with the help of a pump. Materials of much smaller particle size and hence very high surface area can be used for column packing. This allows a much better separation of the components in the mixture. The most common application of HPLC is in analysis [181].

Ion-Exchange Chromatography (IEC)

IEC is often applied to the separation of acidic or basic samples, whose charge varies with pH. In this technique, the separation occurs through exchange or interchange of ions between the sample solution and the solid stationary phase. Primarily the ion exchange chromatography is regulated by electrostatic interactions between the ions being exchanged (the mobile ions in the sample) and the fixed ions that are attached covalently to the molecular lattice of the ion-exchanger.

Affinity Chromatography

This technique is largely used for isolation and purification of biological materials like, proteins, enzymes, antibodies, antigens, viruses, and intact cells by their reactivity with specific immobilized substances. This chromatographic technique is based on the principle that the molecule to be purified can form a selective but reversible interaction with another molecular species immobilized on a suitable chromatographic support. When the chromatographic support is a membrane, it is called as affinity membrane chromatography. Microporous/macroporous membranes containing functional ligands attached to their inner pore surface are used as adsorbents. Affinity membrane chromatography is a promising large-scale separation process for the isolation, purification, and recovery of proteins and enzymes [182–184]. Immobilized-metal affinity chromatography is relatively a new and advanced technique appropriate for protein purification [185]. Feng et al. [186] have synthesized a new immobilized-metal affinity chromatography adsorbent with paramagnetism for the separation and purification of protein. Separation through affinity chromatography can be performed by two different techniques, column chromatography and batch methods, of which batch method is usually applied for large-scale preparative separations. Affinity chromatography in batch mode is becoming more popular with the growth of biotechnology industry.

Size Exclusion Chromatography (SEC)

Separation and retention in SEC are determined by the hydrodynamic diameter of the solute molecule relative to the size of the pores of the column packing. Thus, in SEC, the solutes are eluted according to decreasing molecular size and maximum available volume for separation is equal to the total pore volume of the packing medium. This technique is popular for the purification of protein molecules [187, 188] because of high recovery and its ability to remove undesirable aggregates (dimers, oligomers, etc.) from protein products [189]. Technical improvement of SEC to high-performance SEC (HPSEC) led to more rapid separation and increased resolution [190].

Reversed Phase Chromatography (RPC)

This is the most popular liquid chromatographic technique of separation. In this technique, an aqueous/organic solvent mixture is commonly used as the mobile phase and a high surface area nonpolar solid (usually an alkyl-bonded silica packing) is used as the stationary phase.

Chiral Chromatography

This chromatographic technique is primarily used for separating stereoisomers, that is, chiral compounds.

Thin-Layer Chromatography (TLC)

This is a planar chromatography in which the mobile phase moves through the stationary phase of a thin layer of adsorbent like silica gel, alumina, or cellulose, on a flat and inert substrate. The driving force of the solvent system and the retarding action of the stationary phase are responsible for the separation in TLC. Compared to other planar chromatographic techniques, it has the advantage of faster runs, better separations, and the choice of variety of adsorbents. High-performance TLC has also been developed for the better resolution. TLC is a laboratory separation technique that is used for the separation of organic compounds in solution.

The chromatographic separations can be performed in batch as well as in continuous mode [191]. Among all the above chromatographic techniques, different forms of column chromatography in continuous mode (simulated moving-bed chromatography, SMBC) are the most widely used technique for industrial separation. Simulated moving-bed chromatography (SMBC) is the most widely used technique as it requires less solvent for product elution and would be less cost intensive than for batch chromatography [191].

The continuous simulated moving bed chromatographic reactors have been extensively studied for many different separation schemes arising in the fine chemical, pharmaceutical, biotechnology, and petrochemical industries. Typical successful examples include p-xylene separation from its C8 isomers, n-paraffins from branched and cyclic hydrocarbons, olefins from paraffins, sugar-processing industry, chiral and enantiomers of isoflurane, enflurane, dilute binary gases, and other multicomponent systems.

The operation of a simulated moving bed provides a practically convenient way to mimic true moving bed acquiring all its advantages while eliminating the disadvantages arising due to solids motion. Over the years, a number of improvements in SMB operation modes such as in gas phase, supercritical conditions, and so forth have been made for specific applications. A generic design strategy for an SMBC separation system begins with the knowledge of the equilibrium and kinetic parameters with a view to determine the operating parameters, amount of adsorbent, pressure constraints, column geometry, and so forth for obtaining maximum productivity value. The results provide an assessment of whether the process is reasonable. Continuous models accounting for the mass transfer, heat transfer, pressure drop effects, and so forth are solved providing separation region analysis. An optimization procedure is then implemented to improve separation productivity for a given adsorbent/desorbant consumption, feed flow rates, constant columns number, and arrangement. The optimized solution is then detailed and simulated with a discontinuous model.

A typical case study employs the SMBC for the downstream processing of recombinant proteins using a two-step salt gradient in three-zone open loop continuous countercurrent process for the purification of recombinant streptokinase. Possible process parameters and conditions for efficient separation are obtained through simulations. The model-based design of an SMBC unit requires the knowledge of the properties of the adsorbent phase such as its particle size, density, porosity, pore radius, heat capacity, and so forth. Likewise the model needs the data on the fluid phase components equilibrium adsorption parameters over the adsorbent materials such as its maximum loading capacity, adsorption equilibrium constants, and isosteric heat of adsorption of the components. Additionally, knowledge of Henry constants at different temperatures and mass transfer parameters (pore diffusion, axial dispersion, and external mass transfer coefficient) and heat transfer (heat conductivity, component fluid heat capacities, and Biot number) parameters is essential.

Simulated moving-bed chromatography (SMBC) is the most popular method for industrial separation of optical isomers [192–194], where a number of columns are connected in series with inlet/outlet lines connected between the columns. Recently, a multicolumn continuous chromatographic separation method named as "VARICOL" has been developed by Ludemann-Hombourger and his coworkers [195, 196] which is based on the principle of asynchronous shift of the inlet/outlet lines in a multicolumn system on a recycle loop. The newly developed method "Varicol" is proposed to be more efficient than the traditional simulated moving-bed technique. Another form of chiral resolution process for the separation of racemic mixture of difluoromethylornithine

(DFMO HCl) in industrial scale has been proposed by Perrin et al. [197] involving a multicolumn continuous enantioselective chromatographic process coupled with enantioselective crystallization process. The other chromatographic techniques used for chiral separation are high-performance liquid chromatography (HPLC), gas chromatography (GC), supercritical fluid chromatography (SFC), thin-layer chromatography (TLC), and capillary electrochromatography (CEC).

The SMBC technique involving various separation mechanisms (adsorption, ion exchange, and so forth) is also being successfully employed for the separation of p-xylene isomers [198] and in the separation of glucose and fructose [199–201]. The new challenge for the SMB technology is its application to the separation and purification of biomolecules [202]. Examples of products that are considered for SMBC separation and purification are proteins [203, 204], amino acids [205], antibodies [206], nucleosides [207], and plasmid DNA [208].

Du et al. [209] have developed a system for low-speed rotary countercurrent chromatography by utilizing a convoluted multilayer helical tubing to be used for industrial separation. High-performance centrifugal partition chromatography (HPCPC) is also a practical and suitable method for the separation of biomolecules such as proteins, enzymes, and so forth, particularly on the preparative scale [210].

Electroosmosis

Electroosmosis, also called electroendosmosis, is the motion of polar liquid through a porous membrane or any other porous structure under the influence of an applied electric field. If a solution is separated by a porous diaphragm and an e.m.f. is applied between the electrodes placed on each side of the diaphragm, there will be a flow of liquid from one side to the other. The movement of liquid is also known as electroosmotic flow. A porous diaphragm behaves as a mass of small capillaries. Glass capillaries can also be used for the electroosmotic flow to be observed. In each case, the charged layer attached to the solid cannot move and so the diffused layer in the liquid phase together with the liquid moves under the influence of an electric field. The direction of the electroosmotic flow depends on the diffuse part of the double layer. In moderately pure water, most solids acquire a negative charge so that the diffuse layer has a resultant positive charge, and hence, the flow of water is generally towards the cathode.

Electroosmotic separation technique has industrial importance for separation of water from colloidal suspensions as it consumes less energy than the conventional technique like evaporation. The most recent development in this technology has involved enhanced mixing efficiency by introducing a nonuniform zeta potential, and the enhancement in mixing was experimentally observed by Herr et al. [211]. In 1930, Bartow and Jebens [212] reported water purification by Electroosmosis through a diaphragm. The water so purified was claimed to be equivalent to distilled water and could be obtained at a much lower cost. Electroosmotic separation technique was applied along with vacuum separation technique for the separation of iron oxide ultrafines [213]. The cationic surfactant CTAB was used to maintain the zeta potential in order to achieve the electroosmotic separation of the uncharged ultrafines from iron oxide slurry. A sludge (water treatment process sludge) treatment method by electroosmotic dewatering has been proposed by Buijs et al. [214]. Buijs and coworkers have performed electroosmotic dewatering of commercial sludge on pilot plant scale and on real process scale. The energy consumption for dewatering was found to be within the range of 20–40 kJ kg^{-1}. The removal of water from food suspensions is of great interest in food technology. Al-Asheh et al. [215] reported a direct current Electroosmosis dewatering technique to concentrate tomato to the conventional paste suspension. The process was claimed to save 70% of energy as compared to the evaporation technique. The electroosmotic technique has been used by several researchers for removing water soluble organics from soil for soil remediation purpose [216], dewatering of filter cakes of activated sludge [217], and so forth.

Hybrid Separation Techniques

In addition to the separation techniques described above, several forms of hybrid separation processes have been developed and applied to achieve the desired separation in dilute liquid streams. These hybrid processes usually involve two or more than two different separation techniques. Sarangi and Pattanaik have developed such a separation process comprising of reverse osmosis (RO) and solvent extraction techniques for recovering copper from dilute solutions [218]. An integrative membrane coagulation-adsorption bioreactor was developed by Tian et al. [219] for the removal of organic matter in drinking water treatment. Many novel separation processes involving electrophoresis and other separation techniques like solvent extraction, chromatography, isoelectric focusing are developed for the separation of ionic biomolecules as well as inorganic ions [180, 220]. Spoor et al. [221] reported pilot scale deionization of galvanic nickel solution using a hybrid ion-exchange/electrodialysis system. Decolourization of reactive dyes from an aqueous solution by combined coagulation/ micellar-enhanced ultrafiltration process has been reported by Ahmad and Puasa [222]. Den and Wang [223] suggested a hybrid separation method for treatment of brackish water. The method consists of pretreatment of the brackish water by electrocoagulation (for removing silica) followed by reverse osmosis. According to Den and Wang, pretreatment of brackish water by electrocoagulation prevents the fouling of RO membrane and hence increases the membrane life. Nataraj et al. have developed hybrid separation techniques by combining nanofiltration (NF) with reverse osmosis (RO), microfiltration (MF) with electrodialysis (ED) and took the techniques upto pilot-scale level for water/wastewater treatment. Nataraj et al. [224] have reported a pilot-scale skid-mounted system comprising of nanofiltration (NF) and reverse osmosis membrane processes for the treatment of distillery wastewater. Commercial NF membrane and thin-film composite (TFC) polyamide RO membrane in spiral wound configuration was used in the process. The same NF-RO process was successfully used for removing dye and salts from simulated water in pilot scale by Nataraj et al. [225]. Pilot plant of a hybrid microfiltration (MF) and electrodialysis (ED) system was also designed, constructed, and employed successfully by Nataraj et al. [226] for the treatment of paper industry wastewater. In this hybrid MF-ED process, Nataraj et al. have used microfiltration module of ceramic membrane as a pretreatment step for the electrodialysis pilot-scale unit operation. The electrodialysis pilot-scale unit designed by them consisted of membrane stack of a series of cation exchange and anion exchange membranes. The hybrid MF-ED process could recover 80% of the wastewater.

DISCUSSIONS AND FUTURE NEEDS

Developments of separation and purification technology for processing of dilute liquid streams have been reviewed with special focus on adsorption, membrane-based techniques, chromatography, and electrokinetic-based separation techniques (Electroosmosis, electrophoresis).

Because of the vast applicability of adsorption technology, it has grown extensively over the past years, and still new/modified adsorbents are being added into the list. Also in the recent years, a lot of novel adsorption processes have been developed for enhanced separation and purification of the processed streams. Despite the development of a large number of new/modified adsorbents, very few of them are set for their practical application in industrial processed streams. Successful implementation of adsorbent and adsorption necessitates reasonably good understanding of the fundamentals of the process. Synthesis of advanced materials with high chemical and thermal stability, cost effectiveness of the process, improved predictive models for mass transfer, adsorption, equilibrium, and other physical data is the challenges for further growth of this technology in future.

The extensive literature on membrane separation indicates that the use of membranes in separation and purification processes is growing steadily, and it is believed that membrane separation can play an important role in reducing the environmental impact largely. A wide range of membrane materials has been investigated by the researchers to achieve enhanced and quality separation. Nevertheless, the major constraint for limited application of this technology is high scaleup costs for large-scale applications. It is required to develop robust functional tailored materials with high mechanical and thermal stability, enhanced permeability and selectivity, and economic process scaleup for extensive and versatile application of the technique and to meet the future needs.

Chromatographic separation in batch mode is considered to be very expensive technique. Nevertheless, the continuous chromatography have shown considerable advantages over the batch mode as it better utilizes the adsorbent materials, reduce solvent consumption, and increase productivity. Simulated moving bed chromatography is the most popular among all existing separation techniques for separation of chiral compounds in pharmaceutical industries and separation of biomolecules in bio-chemical and biomedical applications. Among a vast range of existing materials, only very few can practically serve as the perfect materials for specific separation to be used as stationary phase in industrial applications. Hence, there is a necessity to develop/identify materials suitable for stationary phase and mobile phase. Also there is urgent need to develop flexible, fast scaleup, and economically viable process for enhanced separation.

Notwithstanding the differences in the strategies for separation, the research needs for further improvements in the separation techniques can be covered under the common categories as [227] the following.

- Synthesis of new advanced materials (adsorbents, membrane materials, specialty materials, functional polymers, etc.) improves performance.
- Their physicochemical data bases (thermodynamics and kinetics parameters).
- Understanding the role and effects of uncertainties in proposed mechanism (such as nucleation, growth, surface interaction, transport, etc.).
- Understanding the dynamics of exchange processes at the interface.
- Lack of direct measurements of variables of interest (e.g., in situ sampling, analytical and flow visualization).
- Development of sensors and other analytical measurement instruments.
- Development of better predictive modeling tools covering length-time scales of interest.
- Use of models for product quality control and optimization of process systems
- Hybridization: combined separation operations or reaction and separation.
- Molecular recognition as a basis for separation.
- Fundamental understanding, new equipment, and test facilities.

ACKNOWLEDGMENTS

The authors wish to acknowledge the financial support by Council of Scientific and Industrial Research, India.

REFERENCES

1. C. D. Chudasama, J. Sebastian, and R. V. Jasra, "Pore-size engineering of zeolite A for the size/shape selective molecular separation," Industrial and Engineering Chemistry Research, vol. 44, no. 6, pp. 1780–1786, 2005.

2. H. Herden, W.-D. Einicke, M. Jusek, U. Messow, and R. Schöllner, "Adsorption studies of n-Olefin/n-paraffin mixtures on X- and Y-zeolites. I. Comparison of liquid phase and vapor phase adsorption of hexane-1 and n-hexane on NaX-zeolite," Journal of Colloid And Interface Science, vol. 97, no. 2, pp. 559–564, 1984.

3. Y. L. Cheng and T. Y. Lee, "Separation of fructose and glucose mixture by zeolite Y," Biotechnology and Bioengineering, vol. 40, no. 4, pp. 498–504, 1992. · ·

4. A. Methivier, "Influence of oxygenated contaminants on the separation of C_8 aromatics by adsorption on faujasite zeolites," Industrial and Engineering Chemistry Research, vol. 37, no. 2, pp. 604–608, 1998.

5. Z. Guo, S. Zheng, and Z. Zheng, "Separation of p-chloronitrobenzene and o-chloronitrobenzene by selective adsorption using silicalite-1 zeolite," Chemical Engineering Journal, vol. 155, no. 3, pp. 654–659, 2009. · ·

6. S. Al-Asheh, F. Banat, and N. Al-Lagtah, "Separation of ethanol-water mixtures using molecular sieves and biobased adsorbents," Chemical Engineering Research and Design, vol. 82, no. 7, pp. 855–864, 2004. · ·

7. R. B. Eldridge, "Olefin/paraffin separation technology: a review," Industrial and Engineering Chemistry Research, vol. 32, no. 10, pp. 2208–2212, 1993.

8. M. Maes, L. Alaerts, F. Vermoortele et al., "Separation of C5-hydrocarbons on microporous materials: complementary performance of MOFs and zeolites," Journal of the American Chemical Society, vol. 132, no. 7, pp. 2284–2292, 2010. · ·

9. H. Herden, W. D. Einicke, and R. Schöllner, "Adsorption of n-hexane/n-olefin mixtures by NaX zeolites from liquid solution," Journal of Colloid And Interface Science, vol. 79, no. 1, pp. 280–283, 1981.

10. H. Herden, W. D. Einicke, U. Messow, K. Quitzsch, and R. Schöllner, "Adsorption studies ofn-olefin/n-paraffin mixtures on X- and Y-zeolites. II. Adsorption of tetradecene-1/n-dodecane mixtures on modified X- and Y-zeolites," Journal of Colloid And Interface Science, vol. 97, no. 2, pp. 565–573, 1984.

11. R. T. Yang and E. S. Kikkinides, "New sorbents for olefin/paraffin separations by adsorption via ϖ-complexation," AIChE Journal, vol. 41, no. 3, pp. 509–517, 1995.

12. A. van Miltenburg, W. Zhu, F. Kapteijn, and J. A. Moulijn, "Adsorptive separation of light olefin/paraffin mixtures," Chemical Engineering Research and Design, vol. 84, no. 5A, pp. 350–354, 2006. · ·

13. J. Padin, R. T. Yang, and C. L. Munson, "New sorbents for olefin/paraffin separations and olefin purification for C_4 hydrocarbons," Industrial and Engineering Chemistry Research, vol. 38, no. 10, pp. 3614–3621, 1999.

14. A. J. Hernández-Maldonado and R. T. Yang, "Desulfurization of commercial liquid fuels by selective adsorption via ϖ-complexation with Cu(I)-Y zeolite," Industrial and Engineering Chemistry Research, vol. 42, no. 13, pp. 3103–3110, 2003.

15. H. Yang and B. Tatarchuk, "Novel-doped zinc oxide sorbents for low temperature regenerable desulfurization applications," AIChE Journal, vol. 56, no. 11, pp. 2898–2904, 2010. · ·

16. A. Takahashi and R. T. Yang, "New adsorbents for purification: selective removal of aromatics," AIChE Journal, vol. 48, no. 7, pp. 1457–1468, 2002. · ·

17. J. W. Lee, H. C. Park, and H. Moon, "Adsorption and desorption of cephalosporin C on nonionic polymeric sorbents," Separation and Purification Technology, vol. 12, no. 1, pp. 1–11, 1997. · ·

18. D. Y. Cha, "Simple model of irreversible ion-exchange kinetics for sorption of antibiotics from fermentation broth," Reactive Polymers, Ion Exchangers, Sorbents, vol. 5, no. 3, pp. 269–279, 1987.

19. M. H. L. Ribeiro and I. A. C. Ribeiro, "Recovery of erythromycin from fermentation broth by adsorption onto neutral and ion-exchange resins," Separation and Purification Technology, vol. 45, no. 3, pp. 232–239, 2005. · ·

20. M. Dutta, N. N. Dutta, and K. G. Bhattacharya, "Aqueous phase adsorption of certain beta-lactam antibiotics onto polymeric resins and activated carbon," Separation and Purification Technology, vol. 16, no. 3, pp. 213–224, 1999. · ·

21. W. S. Adriano, V. Veredas, C. C. Santana, and L. R. B. Gonçalves, "Adsorption of amoxicillin on chitosan beads: kinetics, equilibrium and validation of finite bath models," Biochemical Engineering Journal, vol. 27, no. 2, pp. 132–137, 2005. · ·

22. S. M. Solberg and C. C. Landry, "Adsorption of DNA into mesoporous silica," Journal of Physical Chemistry B, vol. 110, no. 31, pp. 15261–15268, 2006. · ·

23. B. Saoudi, N. Jammul, M. M. Chehimi, G. P. McCarthy, and S. P. Armes, "Adsorption of DNA onto polypyrrole-silica nanocomposites," Journal of Colloid and Interface Science, vol. 192, no. 1, pp. 269–273, 1997. · ·

24. P. Cai, Q. Huang, and X. Zhang, "Microcalorimetric studies of the effects of $MgCl_2$ concentrations and pH on the adsorption of DNA on montmorillonite, kaolinite and goethite," Applied Clay Science, vol. 32, no. 1-2, pp. 147–152, 2006. · ·

25. X. Zhao and J. K. Johnson, "Simulation of adsorption of DNA on carbon nanotubes," Journal of the American Chemical Society, vol. 129, no. 34, pp. 10438–10445, 2007.

26. X. Zhou, B. Xue, S. Bai, and Y. Sun, "Macroporous polymeric ion exchanger of high capacity for protein adsorption," Biochemical Engineering Journal, vol. 11, no. 1, pp. 13–17, 2002. · ·

27. Z. Ma, Y. Guan, and H. Liu, "Superparamagnetic silica nanoparticles with immobilized metal affinity ligands for protein adsorption," Journal of Magnetism and Magnetic Materials, vol. 301, no. 2, pp. 469–477, 2006. · ·

28. K. Yao, S. Shen, J. Yun, L. Wang, F. Chen, and X. Yu, "Protein adsorption in supermacroporous cryogels with embedded nanoparticles," Biochemical Engineering Journal, vol. 36, no. 2, pp. 139–146, 2007. · ·

29. X. Zhang, R. Bai, and Y. W. Tong, "Selective adsorption behaviors of proteins on polypyrrole-based adsorbents," Separation and Purification Technology, vol. 52, no. 1, pp. 161–169, 2006. · ·

30. X. Diao, Y. Wang, J. Zhao, and S. Zhu, "Effect of pore-size of mesoporous SBA-15 on adsorption of bovine serum albumin and lysozyme protein," Chinese Journal of Chemical Engineering, vol. 18, no. 3, pp. 493–499, 2010. · ·

31. Y. S. Chen, C. S. Chang, and S. Y. Suen, "Protein adsorption separation using glass fiber membranes modified with short-chain organosilicon derivatives," Journal of Membrane Science, vol. 305, no. 1-2, pp. 125–135, 2007. · ·

32. Z. Feng, Z. Shao, J. Yao, Y. Huang, and X. Chen, "Protein adsorption and separation with chitosan-based amphoteric membranes," Polymer, vol. 50, no. 5, pp. 1257–1263, 2009. · ·

33. J. S. Becker, O. R. T. Thomas, and M. Franzreb, "Protein separation with magnetic adsorbents in micellar aqueous two-phase systems," Separation and Purification Technology, vol. 65, no. 1, pp. 46–53, 2009. · ·

34. J. Qu, "Research progress of novel adsorption processes in water purification: a review,"Journal of Environmental Sciences, vol. 20, no. 1, pp. 1–13, 2008. · ·

35. H. Sontheimer, J. C. Crittenden, and R. S. Summers, Activated Carbon for Water Treatment, DVGW-Forschungsstelle, Karlsruhe, Germany, 2nd edition, 1988.

36. N. H. Phan, S. Rio, C. Faur, L. Le Coq, P. Le Cloirec, and T. H. Nguyen, "Production of fibrous activated carbons from natural cellulose (jute, coconut) fibers for water treatment applications," Carbon, vol. 44, no. 12, pp. 2569–2577, 2006. · ·

37. S. M. Maliyekkal, S. Shukla, L. Philip, and I. M. Nambi, "Enhanced fluoride removal from drinking water by magnesia-amended activated alumina granules," Chemical Engineering Journal, vol. 140, no. 1–3, pp. 183–192, 2008. · ·

38. A. Bhatnagar, E. Kumar, and M. Sillanpää, "Nitrate removal from water by nano-alumina: characterization and sorption studies," Chemical Engineering Journal, vol. 163, no. 3, pp. 317–323, 2010. · ·

39. B. Qu, J. Zhou, X. Xiang, C. Zheng, H. Zhao, and X. Zhou, "Adsorption behavior of azo dye C. I. Acid Red 14 in aqueous solution on surface soils," Journal of Environmental Sciences, vol. 20, no. 6, pp. 704–709, 2008. · ·

40. S. Maity, Study on naturally occurring minerals for the removal of arsenic from groundwater, Ph.D. thesis, Jadavpur University, Kolkata, India, 2004.

41. S. Wang and Y. Peng, "Natural zeolites as effective adsorbents in water and wastewater treatment," Chemical Engineering Journal, vol. 156, no. 1, pp. 11–24, 2010. · ·

42. K. H. Goh, T. T. Lim, and Z. Dong, "Application of layered double hydroxides for removal of oxyanions: a review," Water Research, vol. 42, no. 6-7, pp. 1343–1368, 2008. · ·

43. S. Mandal, S. Mayadevi, and B. D. Kulkarni, "Adsorption of aqueous selenite [Se(IV)] species on synthetic layered double hydroxide materials," Industrial and Engineering Chemistry Research, vol. 48, no. 17, pp. 7893–7898, 2009. · ·

44. K. S. Hui, C. Y. H. Chao, and S. C. Kot, "Removal of mixed heavy metal ions in wastewater by zeolite 4A and residual products from recycled coal fly ash," Journal of Hazardous Materials, vol. 127, no. 1–3, pp. 89–101, 2005. · ·

45. C. R. Oliveira and J. Rubio, "New basis for adsorption of ionic pollutants onto modified zeolites," Minerals Engineering, vol. 20, no. 6, pp. 552–558, 2007. · ·

46. M. Carmona, A. D. Lucas, J. L. Valverde, B. Velasco, and J. F. Rodríguez, "Combined adsorption and ion exchange equilibrium of phenol on Amberlite IRA-420," Chemical Engineering Journal, vol. 117, no. 2, pp. 155–160, 2006. · ·

47. A. Nilchi, A. A. Babalou, R. Rafice, and H. S. Kalal, "Adsorption properties of amidoxime resins for separation of metal ions from aqueous systems," Reactive and Functional Polymers, vol. 68, no. 12, pp. 1663–1668, 2008. · ·

48. I. Abe, K. Hayashi, and H. Tatsumoto, "The relation between activated carbon adsorption and water quality indexes," Water Research, vol. 19, no. 9, pp. 1191–1193, 1985. · ·

49. I. Ali, "The quest for active carbon adsorbent substitutes: inexpensive adsorbents for toxic metal ions removal from wastewater," Separation and Purification Reviews, vol. 39, no. 3-4, pp. 95–171, 2010. · ·

50. B. Li, Z. Lei, and Z. Huang, "Surface-treated activated carbon for removal of aromatic compounds from water," Chemical

Engineering and Technology, vol. 32, no. 5, pp. 763–770, 2009. · ·

51.   J. Zhu, H. Liang, J. Fang, J. Zhu, and B. Shi, "Characterization of chlorinated tire-derived mesoporous activated carbon for adsorptive removal of toluene," CLEAN—Soil, Air, Water, vol. 39, no. 6, pp. 557–565, 2011. ·

52.   R. U. Edgehill and G. Q. Lu, "Adsorption characteristics of carbonized bark for phenol and pentachlorophenol," Journal of Chemical Technology and Biotechnology, vol. 71, no. 1, pp. 27–34, 1998. · ·

53.   E. Ayranci and N. Hoda, "Adsorption kinetics and isotherms of pesticides onto activated carbon-cloth," Chemosphere, vol. 60, no. 11, pp. 1600–1607, 2005. · ·

54.   E. Ayranci and N. Hoda, "Adsorption of bentazon and propanil from aqueous solutions at the high area activated carbon-cloth," Chemosphere, vol. 57, no. 8, pp. 755–762, 2004. · ·

55.   A. R. Binupriya, M. Sathishkumar, S. H. Jung, S. H. Song, and S. I. Yun, "Bokbunja wine industry waste as precursor material for carbonization and its utilization for the removal of procion red MX-5B from aqueous solutions," CLEAN—Soil, Air, Water, vol. 36, no. 10-11, pp. 879–886, 2008. · ·

56.   N. Thinakaran, P. Baskaralingam, K. V. Thiruvengada Ravi, P. Panneerselvam, and S. Sivanesan, "Adsorptive removal of acid blue 15: equilibrium and kinetic study," CLEAN—Soil, Air, Water, vol. 36, no. 9, pp. 798–804, 2008. · ·

57.   T. Calvete, E. C. Lima, N. F. Cardoso, S. L. P. Dias, and E. S. Ribeiro, "Removal of brilliant green dye from aqueous solutions using home made activated carbons," CLEAN—Soil, Air, Water, vol. 38, no. 5-6, pp. 521–532, 2010. · ·

58.   S. Sinha, K. Pandey, D. Mohan, and K. P. Singh, "Removal of fluoride from aqueous solutions by Eichhornia crassipes biomass and its carbonized form," Industrial and Engineering Chemistry Research, vol. 42, no. 26, pp. 6911–6918, 2003.

59.   X. L. Long, H. Cheng, Z. L. Xin, W. D. Xiao, W. Li, and W. K. Yuan, "Adsorption of ammonia on activated carbon from aqueous solutions," Environmental Progress, vol. 27, no. 2, pp. 225–233, 2008. · ·

60. Y. H. Chang, K. H. Hsieh, and F. C. Chang, "Removal of Hg^{2+} from aqueous solution using a novel composite carbon adsorbent," Journal of Applied Polymer Science, vol. 112, no. 4, pp. 2445–2454, 2009. · ·

61. C. Selomulya, V. Meeyoo, and R. Amal, "Mechanisms of Cr(VI) removal from water by various types of activated carbons," Journal of Chemical Technology and Biotechnology, vol. 74, no. 2, pp. 111–122, 1999. · ·

62. J. P. Chen, S. Wu, and K. H. Chong, "Surface modification of a granular activated carbon by citric acid for enhancement of copper adsorption," Carbon, vol. 41, no. 10, pp. 1979–1986, 2003. · ·

63. M. A. K. M. Hanafiah and W. S. W. Ngah, "Preparation, characterization and adsorption mechanism of Cu(II) onto protonated rubber leaf powder," CLEAN—Soil, Air, Water, vol. 37, no. 9, pp. 696–703, 2009. · ·

64. K. A. Krishnan, A. Sheela, and T. S. Anirudhan, "Kinetic and equilibrium modeling of liquid-phase adsorption of lead and lead chelates on activated carbons," Journal of Chemical Technology and Biotechnology, vol. 78, no. 6, pp. 642–653, 2003. · ·

65. G. Crini, "Recent developments in polysaccharide-based materials used as adsorbents in wastewater treatment," Progress in Polymer Science, vol. 30, no. 1, pp. 38–70, 2005. · ·

66. G. Crini, "Non-conventional low-cost adsorbents for dye removal: a review," Bioresource Technology, vol. 97, no. 9, pp. 1061–1085, 2006. · ·

67. C. Jeon, I. W. Nah, and K. Y. Hwang, "Adsorption of heavy metals using magnetically modified alginic acid," Hydrometallurgy, vol. 86, no. 3-4, pp. 140–146, 2007. · ·

68. J. Kumar, C. Balomajumder, and P. Mondal, "Application of agro-based biomasses for zinc removal from wastewater—a review," CLEAN—Soil, Air, Water, vol. 39, no. 7, pp. 641–652, 2011. ·

69. C. Zevenbergen, L. P. van Reeuwijk, G. Frapporti, R. J. Louws, and R. D. Schuiling, "A simple method for defluoridation of drinking water at village level by adsorption on Ando soil in Kenya," Science of the Total Environment, vol. 188, no. 2-3, pp. 225–232, 1996. · ·

70. W. Jiang, S. Zhang, X. Q. Shan, M. Feng, Y. G. Zhu, and R. G. McLaren, "Adsorption of arsenate on soils. Part 1: laboratory batch experiments using 16 Chinese soils with different physiochemical properties," Environmental Pollution, vol. 138, no. 2, pp. 278–284, 2005. · ·

71. S. K. Maji, A. Pal, and T. Pal, "Arsenic removal from real-life groundwater by adsorption on laterite soil," Journal of Hazardous Materials, vol. 151, no. 2-3, pp. 811–820, 2008. · ·

72. J. L. Dai, M. Zhang, and Y. G. Zhu, "Adsorption and desorption of iodine by various Chinese soils I. Iodate," Environment International, vol. 30, no. 4, pp. 525–530, 2004. · ·

73. H. Q. Hu, J. Z. He, X. Y. Li, and F. Liu, "Effect of several organic acids on phosphate adsorption by variable charge soils of central China," Environment International, vol. 26, no. 5-6, pp. 353–358, 2001. · ·

74. P. R. Shukla, S. Wang, H. M. Ang, and M. O. Tadé, "Synthesis, characterisation, and adsorption evaluation of carbon-natural-zeolite composites," Advanced Powder Technology, vol. 20, no. 3, pp. 245–250, 2009. ·

75. Y. H. Xu, T. Nakajima, and A. Ohki, "Adsorption and removal of arsenic(V) from drinking water by aluminum-loaded Shirasu-zeolite," Journal of Hazardous Materials, vol. 92, no. 3, pp. 275–287, 2002. · ·

76. X. Ren, C. Chen, M. Nagatsu, and X. Wang, "Carbon nanotubes as adsorbents in environmental pollution management: a review," Chemical Engineering Journal, vol. 170, pp. 395–410, 2011. · ·

77. Y. H. Li, S. Wang, A. Cao et al., "Adsorption of fluoride from water by amorphous alumina supported on carbon nanotubes," Chemical Physics Letters, vol. 350, no. 5-6, pp. 412–416, 2001. · ·

78. Y. S. Ok, J. E. Yang, Y. S. Zhang, S. J. Kim, and D. Y. Chung, "Heavy metal adsorption by a formulated zeolite-Portland cement mixture," Journal of Hazardous Materials, vol. 147, no. 1-2, pp. 91–96, 2007. · ·

79. C. D. Johnson and F. Worrall, "Novel granular materials with microcrystalline active surfaces-Waste water treatment applications of zeolite/vermiculite composites," Water Research, vol. 41, no. 10, pp. 2229–2235, 2007. · ·

80. S. Mandal and S. Mayadevi, "Cellulose supported layered double hydroxides for the adsorption of fluoride from aqueous solution," Chemosphere, vol. 72, no. 6, pp. 995–998, 2008. · ·

81. J. Hao, M. J. Han, C. Wang, and X. Meng, "Enhanced removal of arsenite from water by a mesoporous hybrid material—thiol-functionalized silica coated activated alumina,"Microporous and Mesoporous Materials, vol. 124, no. 1–3, pp. 1–7, 2009. · ·

82. Z. Li, S. Deng, G. Yu, J. Huang, and V. C. Lim, "As(V) and As(III) removal from water by a Ce-Ti oxide adsorbent: behavior and mechanism," Chemical Engineering Journal, vol. 161, no. 1-2, pp. 106–113, 2010. · ·

83. F. P. Cuperus, "Membrane processes in agro-food state-of-the-art and new opportunities,"Separation and Purification Technology, vol. 14, no. 1–3, pp. 233–239, 1998. · ·

84. B. Girard and L. R. Fukumoto, "Membrane processing of fruit juices and beverages: a review," Critical Reviews in Food Science and Nutrition, vol. 40, no. 2, pp. 91–157, 2000.

85. R. van Reis and A. Zydney, "Membrane separations in biotechnology," Current Opinion in Biotechnology, vol. 12, no. 2, pp. 208–211, 2001. · ·

86. C. Charcosset, "Membrane processes in biotechnology: an overview," Biotechnology Advances, vol. 24, no. 5, pp. 482–492, 2006. · ·

87. F. T. Awadalla and A. Kumar, "Opportunities for membrane technologies in the treatment of mining and mineral process streams and effluents," Separation Science and Technology, vol. 29, no. 10, pp. 1231–1249, 1994.

88. W. Doyen, "Latest developments in ultrafiltration for large-scale drinking water applications," Desalination, vol. 113, no. 2-3, pp. 165–177, 1997.

89. A. Bottino, C. Capannelli, A. Del Borghi, M. Colombino, and O. Conio, "Water treatment for drinking purpose: ceramic microfiltration application," Desalination, vol. 141, no. 1, pp. 75–79, 2001. · ·

90. P. Lipp, M. Witte, G. Baldauf, and A. A. Povorov, "Treatment of reservoir water with a backwashable MF/UF spiral wound membrane," Desalination, vol. 179, no. 1–3, pp. 83–94, 2005. · ·

91. J. Mueller, Y. Cen, and R. H. Davis, "Crossflow microfiltration of oily water," Journal of Membrane Science, vol. 129, no. 2, pp. 221–235, 1997. · ·

92. H. Liang, W. Gong, and G. Li, "Performance evaluation of water treatment ultrafiltration pilot plants treating algae-rich reservoir water," Desalination, vol. 221, no. 1–3, pp. 345–350, 2008. · ·

93. S. S. Madaeni, A. G. Fane, and G. S. Grohmann, "Virus removal from water and wastewater using membranes," Journal of Membrane Science, vol. 102, no. 1–3, pp. 65–75, 1995. · ·

94. L. L. Zhang, D. Yang, Z. J. Zhong, and P. Gu, "Application of hybrid coagulation-microfiltration process for treatment of membrane backwash water from waterworks," Separation and Purification Technology, vol. 62, no. 2, pp. 415–422, 2008. · ·

95. P. Drogui, S. Elmaleh, M. Rumeau, C. Bernard, and A. Rambaud, "Hybride process, microfiltration-electroperoxidation, for water treatment," Journal of Membrane Science, vol. 186, no. 1, pp. 123–132, 2001. · ·

96. B. Schlichter, V. Mavrov, and H. Chmiel, "Study of a hybrid process combining ozonation and microfiltration/ultrafiltration for drinking water production from surface water," Desalination, vol. 168, no. 1–3, pp. 307–317, 2004. · ·

97. F. Malek, J. L. Harris, and F. A. Roddick, "Interrelationship of photooxidation and microfiltration in drinking water treatment," Journal of Membrane Science, vol. 281, no. 1-2, pp. 541–547, 2006. · ·

98. Y. T. Tsai, Y. H. Weng, A. Y. C. Lin, and K. C. Li, "Electro-microfiltration treatment of water containing natural organic matter and inorganic particles," Desalination, vol. 267, no. 2-3, pp. 133–138, 2011. · ·

99. F. V. Kosikowski and V. V. Mistry, "Microfiltration, ultrafiltration, and centrifugation separation and sterilization processes for improving milk and cheese quality," Journal of Dairy Science, vol. 73, pp. 1411–1419, 1990.

100. P. Morin, M. Britten, R. Jiménez-Flores, and Y. Pouliot, "Microfiltration of buttermilk and washed cream buttermilk for concentration of milk fat globule membrane components," Journal of Dairy Science, vol. 90, no. 5, pp. 2132–2140, 2007. · ·

101. R. S. Barhate, R. Subramanian, K. E. Nandini, and H. U. Hebbar, "Processing of honey using polymeric microfiltration and ultrafiltration membranes," Journal of Food Engineering, vol. 60, no. 1, pp. 49–54, 2003. · ·

102. F. Vaillant, A. Millan, M. Dornier, M. Decloux, and M. Reynes, "Strategy for economical optimization of the clarification of pulpy fruit juices using crossflow microfiltration," Journal of Food Engineering, vol. 48, no. 1, pp. 83–90, 2001. · ·

103. S. Mondal, A. Cassano, F. Tasselli, and S. De, "A generalized model for clarification of fruit juice during ultrafiltration under total recycle and batch mode," Journal of Membrane Science, vol. 366, no. 1-2, pp. 295–303, 2011. · ·

104. M. Hamachi, B. B. Gupta, and R. B. Aim, "Ultrafiltration: a means for decolorization of cane sugar solution," Separation and Purification Technology, vol. 30, no. 3, pp. 229–239, 2003. · ·

105. B. K. Nelson and D. M. Barbano, "A microfiltration process to maximize removal of serum proteins from skim milk before cheese making," Journal of Dairy Science, vol. 88, no. 5, pp. 1891–1900, 2005.

106. T. Y. Wu, A. W. Mohammad, J. M. Jahim, and N. Anuar, "Palm oil mill effluent (POME) treatment and bioresources recovery using ultrafiltration membrane: effect of pressure on membrane fouling," Biochemical Engineering Journal, vol. 35, no. 3, pp. 309–317, 2007. · ·

107. M. D. Afonso and R. Bórquez, "Review of the treatment of seafood processing wastewaters and recovery of proteins therein by membrane separation processes—prospects of the ultrafiltration of wastewaters from the fish meal industry," Desalination, vol. 142, no. 1, pp. 29–45, 2002. · ·

108. Y. M. Lo, D. Cao, S. Argin-Soysal, J. Wang, and T. S. Hahm, "Recovery of protein from poultry processing wastewater using membrane ultrafiltration," Bioresource Technology, vol. 96, no. 6, pp. 687–698, 2005. · ·

109. Y. Li, A. Shahbazi, and C. T. Kadzere, "Separation of cells and proteins from fermentation broth using ultrafiltration," Journal of Food Engineering, vol. 75, no. 4, pp. 574–580, 2006. · ·

110. C. W. Cho, D. Y. Lee, and C. W. Kim, "Concentration and purification of soluble pectin from mandarin peels using crossflow microfiltration system," Carbohydrate Polymers, vol. 54, no. 1, pp. 21–26, 2003. · ·

111. V. G. Molina and A. Casañas, "Reverse osmosis, a key technology in combating water scarcity in Spain," Desalination, vol. 250, no. 3, pp. 950–955, 2010. · ·

112. L. Malaeb and G. M. Ayoub, "Reverse osmosis technology for water treatment: state of the art review," Desalination, vol. 267, no. 1, pp. 1–8, 2011. · ·

113. A. Almulla, M. Eid, P. Côté, and J. Coburn, "Developments in high recovery brackish water desalination plants as part of the solution yp water quantity problems," Desalination, vol. 153, no. 1–3, pp. 237–243, 2003. · ·

114. J. C. Schrotter, S. Rapenne, J. Leparc, P. J. Remize, and S. Casas, "Current and emerging developments in desalination with reverse osmosis membrane systems," in Comprehensive Membrane Science and Engineering, E. Drioli and L. Giorno, Eds., chapter 2.03, pp. 35–65, Elsevier, New York, NY, USA, 2010.

115. M. Liu, S. Yu, J. Tao, and C. Gao, "Preparation, structure characteristics and separation properties of thin-film composite polyamide-urethane seawater reverse osmosis membrane," Journal of Membrane Science, vol. 325, no. 2, pp. 947–956, 2008. · ·

116. M. Duan, Z. Wang, J. Xu, J. Wang, and S. Wang, "Influence of hexamethyl phosphoramide on polyamide composite reverse osmosis membrane performance," Separation and Purification Technology, vol. 75, no. 2, pp. 145–155, 2010. · ·

117. J. C. Kruithof, J. C. Schippers, P. C. Kamp, H. C. Folmer, and J. A. M. H. Hofman, "Integrated multi-objective membrane systems for surface water treatment: pretreatment of reverse osmosis by conventional treatment and ultrafiltration," Desalination, vol. 117, no. 1–3, pp. 37–48, 1998. · ·

118. A. M›nif, S. Bouguecha, B. Hamrouni, and M. Dhahbi, "Coupling of membrane processes for brackish water desalination," Desalination, vol. 203, no. 1–3, pp. 331–336, 2007. · ·

119. N. Prihasto, Q. F. Liu, and S. H. Kim, "Pre-treatment strategies for seawater desalination by reverse osmosis system," Desalination, vol. 249, no. 1, pp. 308–316, 2009. · ·

120. K. L. Tu, L. D. Nghiem, and A. R. Chivas, "Boron removal by reverse osmosis membranes in seawater desalination applications," Separation and Purification Technology, vol. 75, no. 2, pp. 87–101, 2010. · ·

121. X. Chai, G. Chen, P. L. Yue, and Y. Mi, "Pilot scale membrane separation of electroplating waste water by reverse osmosis," Journal of Membrane Science, vol. 123, no. 2, pp. 235–242, 1997. · ·

122. P. uda, P. Pospíšil, and J. Tenglerová, "Reverse osmosis in water treatment for boilers,"Desalination, vol. 198, no. 1–3, pp. 41–46, 2006. ·

123. M. Belkacem, S. Bekhti, and K. Bensadok, "Groundwater treatment by reverse osmosis,"Desalination, vol. 206, no. 1–3, pp. 100–106, 2007. · ·

124. M. Melo, H. Schluter, J. Ferreira, R. Magda, A. Júnior, and O. de Aquino, "Advanced performance evaluation of a reverse osmosis treatment for oilfield produced water aiming reuse," Desalination, vol. 250, no. 3, pp. 1016–1018, 2010. · ·

125. M. K. H. Liew, S. Tanaka, and M. Morita, "Separation and purification of lactic acid: fundamental studies on the reverse osmosis down-stream process," Desalination, vol. 101, no. 3, pp. 269–277, 1995.

126. V. V. Goncharuk, D. D. Kucheruk, V. M. Kochkodan, and V. P. Badekha, "Removal of organic substances from aqueous solutions by reagent enhanced reverse osmosis,"Desalination, vol. 143, no. 1, pp. 45–51, 2002. · ·

127. F. Liu, G. Zhang, Q. Meng, and H. Zhang, "Performance of nanofiltration and reverse osmosis membranes in metal effluent treatment," Chinese Journal of Chemical Engineering, vol. 16, no. 3, pp. 441–445, 2008. · ·

128. J. Weißbrodt, M. Manthey, B. Ditgens, G. Laufenberg, and B. Kunz, "Separation of aqueous organic multi-component solutions by reverse osmosis—development of a mass transfer model," Desalination, vol. 133, no. 1, pp. 65–74, 2001. · ·

129. C. H. Liang, "Separation properties of high temperature reverse osmosis membranes for silica removal and boric acid recovery," Journal of Membrane Science, vol. 246, no. 2, pp. 127–135, 2005. · ·

130. M. Sadrzadeh and T. Mohammadi, "Sea water desalination using electrodialysis,"Desalination, vol. 221, no. 1–3, pp. 440–447, 2008. ··

131. M. Sadrzadeh and T. Mohammadi, "Treatment of sea water using electrodialysis: current efficiency evaluation," Desalination, vol. 249, no. 1, pp. 279–285, 2009. ··

132. S. K. Nataraj, K. M. Hosamani, and T. M. Aminabhavi, "Potential application of an electrodialysis pilot plant containing ion-exchange membranes in chromium removal,"Desalination, vol. 217, no. 1–3, pp. 181–190, 2007. ··

133. K. Takahashi, K. Umehara, G. P. T. Cruz, S. Nil, and F. Kawaizumi, "Mutual separation of two monovalent metal ions by multistage electrodialysis," Chemical Engineering Science, vol. 60, no. 3, pp. 727–734, 2005. ··

134. F. Hell, J. Lahnsteiner, H. Frischherz, and G. Baumgartner, "Experience with full-scale electrodialysis for nitrate and hardness removal," Desalination, vol. 117, no. 1–3, pp. 173–180, 1998. ··

135. L. J. Banasiak and A. I. Schäfer, "Removal of boron, fluoride and nitrate by electrodialysis in the presence of organic matter," Journal of Membrane Science, vol. 334, no. 1-2, pp. 101–109, 2009. ··

136. T. Sata, "Studies on ion exchange membranes with permselectivity for specific ions in electrodialysis," Journal of Membrane Science, vol. 93, no. 2, pp. 117–135, 1994. ··

137. M. Demircioglu, N. Kabay, I. Kurucaovali, and E. Ersoz, "Demineralization by electrodialysis (ED)—separation performance and cost comparison for monovalent salts," Desalination, vol. 153, no. 1–3, pp. 329–333, 2003. ··

138. R. Audinos, A. Nassr-Allah, J. R. Alvarez, J. L. Andres, and R. Alvarez, "Electrodialysis in the separation of dilute aqueous solutions of sulfuric and nitric acids," Journal of Membrane Science, vol. 76, no. 2-3, pp. 147–156, 1993. ··

139. G. S. Luo, X. Y. Shan, X. Qi, and Y. C. Lu, "Two-phase electro-electrodialysis for recovery and concentration of citric acid," Separation and Purification Technology, vol. 38, no. 3, pp. 265–271, 2004. ··

140. F. S. Rohman, M. R. Othman, and N. Aziz, "Modeling of batch electrodialysis for hydrochloric acid recovery," Chemical Engineering Journal, vol. 162, no. 2, pp. 466–479, 2010. · ·

141. B. van der Bruggen, A. Koninckx, and C. Vandecasteele, "Separation of monovalent and divalent ions from aqueous solution by electrodialysis and nanofiltration," Water Research, vol. 38, no. 5, pp. 1347–1353, 2004. · ·

142. P. V. Vyas, B. G. Shah, G. S. Trivedi, P. M. Gaur, P. Ray, and S. K. Adhikary, "Separation of inorganic and organic acids from glyoxal by electrodialysis," Desalination, vol. 140, no. 1, pp. 47–54, 2001. · ·

143. H. Habe, T. Fukuoka, D. Kitamoto, and K. Sakaki, "Application of electrodialysis to glycerate recovery from a glycerol containing model solution and culture broth," Journal of Bioscience and Bioengineering, vol. 107, no. 4, pp. 425–428, 2009. ·

144. T. V. Eliseeva, V. A. Shaposhnik, E. V. Krisilova, and A. E. Bukhovets, "Transport of basic amino acids through the ion-exchange membranes and their recovery by electrodialysis,"Desalination, vol. 241, no. 1–3, pp. 86–90, 2009. · ·

145. X. Zhang, W. Lu, H. Ren, and W. Cong, "Recovery of glutamic acid from isoelectric supernatant using electrodialysis," Separation and Purification Technology, vol. 55, no. 2, pp. 274–280, 2007. · ·

146. G. Atungulu, S. Koide, S. Sasaki, and W. Cao, "Ion-exchange membrane mediated electrodialysis of scallop broth: ion, free amino acid and heavy metal profiles," Journal of Food Engineering, vol. 78, no. 4, pp. 1285–1290, 2007. · ·

147. L. Yu, A. Lin, L. Zhang, C. Chen, and W. Jiang, "Application of electrodialysis to the production of Vitamin C," Chemical Engineering Journal, vol. 78, no. 2-3, pp. 153–157, 2000. · ·

148. E. M. van der Ent, P. van Hee, J. T. F. Keurentjes, K. van›t Riet, and A. van der Padt, "Multistage electrodialysis for large-scale separation of racemic mixtures," Journal of Membrane Science, vol. 204, no. 1-2, pp. 173–184, 2002. · ·

149. T. Xu and C. Huang, "Electrodialysis-Based separation technologies: a critical review," AIChE Journal, vol. 54, no. 12, pp. 3147–3159, 2008. · ·

150. L. Firdaous, P. Dhulster, J. Amiot et al., "Concentration and selective separation of bioactive peptides from an alfalfa white protein hydrolysate by electrodialysis with ultrafiltration membranes," Journal of Membrane Science, vol. 329, no. 1-2, pp. 60–67, 2009. · ·

151. L. Bazinet, J. Amiot, J. F. Poulin, D. Labbe, and D. Tremblay, "Process and system for separation of organic charged compounds," International Patent WO 2005/082495A1, 2005.

152. K. Nagai, "Fundamentals and perspectives for pervaporation," in Comprehensive Membrane Science and Engineering, chapter 2.10, pp. 243–271, 2010.

153. L. Hitchens, L. M. Vane, and F. R. Alvarez, "VOC removal from water and surfactant solutions by pervaporation: a pilot study," Separation and Purification Technology, vol. 24, no. 1-2, pp. 67–84, 2001. · ·

154. M. Peng, L. M. Vane, and S. X. Liu, "Recent advances in VOCs removal from water by pervaporation," Journal of Hazardous Materials, vol. 98, no. 1–3, pp. 69–90, 2003. · ·

155. H. O. E. Karlsson and G. Tragardh, "Pervaporation of dilute organic-waters mixtures. A literature review on modelling studies and applications to aroma compound recovery," Journal of Membrane Science, vol. 76, no. 2-3, pp. 121–146, 1993. · ·

156. A. Hasanoğlu, Y. Salt, S. Kele er, S. Özkan, and S. Dinçer, "Pervaporation separation of organics from multicomponent aqueous mixtures," Chemical Engineering and Processing, vol. 46, no. 4, pp. 300–306, 2007. ·

157. B. Smitha, D. Suhanya, S. Sridhar, and M. Ramakrishna, "Separation of organic-organic mixtures by pervaporation—a review," Journal of Membrane Science, vol. 241, no. 1, pp. 1–21, 2004. · ·

158. B. K. Dutta and S. K. Sikdar, "Separation of azeotropic organic liquid mixtures by pervaporation," AIChE Journal, vol. 37, no. 4, pp. 581–588, 1991.

159. A. Hasanoğlu, Y. Salt, S. Kele er, S. Özkan, and S. Dinçer, "Pervaporation separation of ethyl acetate-ethanol binary mixtures using polydimethylsiloxane membranes," Chemical Engineering and Processing, vol. 44, no. 3, pp. 375–381, 2005. ·

160. M. Wessling, U. Werner, and S. T. Hwang, "Pervaporation of aromatic C8-isomers," Journal of Membrane Science, vol. 57, no. 2-3, pp. 257–270, 1991. · ·

161. H. L. Chen, L. G. Wu, J. Tan, and C. L. Zhu, "PVA membrane filled -cyclodextrin for separation of isomeric xylenes by pervaporation," Chemical Engineering Journal, vol. 78, no. 2-3, pp. 159–164, 2000. · ·

162. F. Lipnizki, J. Olsson, and G. Trägårdh, "Scale-up of pervaporation for the recovery of natural aroma compounds in the food industry. Part 1: simulation and performance," Journal of Food Engineering, vol. 54, no. 3, pp. 183–195, 2002. · ·

163. X. Feng and R. Y. M. Huang, "Pervaporation with chitosan membranes. I. Separation of water from ethylene glycol by a chitosan/polysulfone composite membrane," Journal of Membrane Science, vol. 116, no. 1, pp. 67–76, 1996. · ·

164. P. Shao and R. Y. M. Huang, "Polymeric membrane pervaporation," Journal of Membrane Science, vol. 287, no. 2, pp. 162–179, 2007. · ·

165. S. D. Bhat and T. M. Aminabhavi, "Pervaporation separation using sodium alginate and its modified membranes—a review," Separation and Purification Reviews, vol. 36, no. 3, pp. 203–229, 2007. · ·

166. S. G. Adoor, L. S. Manjeshwar, B. V. Kumar Naidu, M. Sairam, and T. M. Aminabhavi, "Poly(vinyl alcohol)/poly(methyl methacrylate) blend membranes for pervaporation separation of water + isopropanol and water + 1,4-dioxane mixtures," Journal of Membrane Science, vol. 280, no. 1-2, pp. 594–602, 2006. ·

167. M. G. Mali, V. T. Magalad, G. S. Gokavi, T. M. Aminabhavi, and K. V. S. N. Raju, "Pervaporation separation of isopropanol-water mixtures using mixed matrix blend membranes of poly(vinyl alcohol)/poly(vinyl pyrrolidone) loaded with phosphomolybdic acid," Journal of Applied Polymer Science, vol. 121, no. 2, pp. 711–719, 2011. ·

168. Y. Wang, L. Yang, G. Luo, and Y. Dai, "Preparation of cellulose acetate membrane filled with metal oxide particles for the pervaporation separation of methanol/methyl tert-butyl ether mixtures," Chemical Engineering Journal, vol. 146, no. 1, pp. 6–10, 2009. · ·

169. S. L. Wee, C. T. Tye, and S. Bhatia, "Membrane separation process-Pervaporation through zeolite membrane," Separation and Purification Technology, vol. 63, no. 3, pp. 500–516, 2008. · ·

170. W. Kujawski, A. Warszawski, W. Ratajczak, T. Porębski, W. Capała, and I. Ostrowska, "Application of pervaporation and adsorption to the phenol removal from wastewater," Separation and Purification Technology, vol. 40, no. 2, pp. 123–132, 2004. ·

171. G. S. Luo, M. Niang, and P. Schaetzel, "Separation of ethyl tert-butyl ether-ethanol by combined pervaporation and distillation," Chemical Engineering Journal, vol. 68, no. 2-3, pp. 139–143, 1997. · ·

172. S. Assabumrungrat, J. Phongpatthanapanich, P. Praserthdam, T. Tagawa, and S. Goto, "Theoretical study on the synthesis of methyl acetate from methanol and acetic acid in pervaporation membrane reactors: effect of continuous-flow modes," Chemical Engineering Journal, vol. 95, no. 1, pp. 57–65, 2003. · ·

173. G. J. S. van der Gulik, R. E. G. Janssen, J. G. Wijers, and J. T. F. Keurentjes, "Hydrodynamics in a ceramic pervaporation membrane reactor for resin production," Chemical Engineering Science, vol. 56, no. 2, pp. 371–379, 2001. · ·

174. B. van der Bruggen, "Pervaporation membrane reactors," in Comprehensive Membrane Science and Engineering, vol. 3, pp. 135–163, 2010.

175. B. G. Park and T. T. Tsotsis, "Models and experiments with pervaporation membrane reactors integrated with an adsorbent system," Chemical Engineering and Processing, vol. 43, no. 9, pp. 1171–1180, 2004. · ·

176. C. Liu, X. Xu, Q. Wang, and J. Chen, "Mathematical model for DNA separation by capillary electrophoresis in entangled polymer solutions," Journal of Chromatography A, vol. 1142, no. 2, pp. 222–230, 2007. · ·

177. A. R. Piergiovanni, "Extraction and separation of water-soluble proteins from different wheat species by acidic capillary electrophoresis," Journal of Agricultural and Food Chemistry, vol. 55, no. 10, pp. 3850–3856, 2007. · ·

178. G. Gübitz and M. G. Schmid, "Chiral separation by chromatographic and electromigration techniques. A review," Biopharmaceutics and Drug Disposition, vol. 22, no. 7-8, pp. 291–336, 2001. ·

179. M. J. Clifton, H. Roux-De Balmann, and V. Sanchez, "Protein separation by continuous-flow electrophoresis in microgravity," AIChE Journal, vol. 42, no. 7, pp. 2069–2078, 1996.

180. S. L. Zhai, G. S. Luo, and J. G. Liu, "Selective recovery of amino acids by aqueous two-phase electrophoresis," Chemical Engineering Journal, vol. 83, no. 1, pp. 55–59, 2001. · ·

181. T. Guerin, A. Astruc, and M. Astruc, "Speciation of arsenic and selenium compounds by HPLC hyphenated to specific detectors: a review of the main separation techniques," Talanta, vol. 50, no. 1, pp. 1–24, 1999. · ·

182. T. Burnouf and M. Radosevich, "Affinity chromatography in the industrial purification of plasma proteins for therapeutic use," Journal of Biochemical and Biophysical Methods, vol. 49, no. 1–3, pp. 575–586, 2001. · ·

183. W. Guo and E. Ruckenstein, "Separation and purification of horseradish peroxidase by membrane affinity chromatography," Journal of Membrane Science, vol. 211, no. 1, pp. 101–111, 2003. · ·

184. M. Yılmaz, G. Bayramoğlu, and M. Y. Arica, "Separation and purification of lysozyme by Reactive Green 19 immobilised membrane affinity chromatography," Food Chemistry, vol. 89, no. 1, pp. 11–18, 2005. · ·

185. V. Gaberc-Porekar and V. Menart, "Perspectives of immobilized-metal affinity chromatography," Journal of Biochemical and Biophysical Methods, vol. 49, no. 1–3, pp. 335–360, 2001. · ·

186. G. Feng, D. Hu, L. Yang, Y. Cui, X. A. Cui, and H. Li, "Immobilized-metal affinity chromatography adsorbent with paramagnetism and its application in purification of histidine-tagged proteins," Separation and Purification Technology, vol. 74, no. 2, pp. 253–260, 2010. · ·

187. E. E. G. Rojas, J. S. Dos Reis Coimbra, L. A. Minim, A. D. G. Zuniga, S. H. Saraiva, and V. P. R. Minim, "Size-exclusion chromatography applied to the purification of whey proteins from

the polymeric and saline phases of aqueous two-phase systems," Process Biochemistry, vol. 39, no. 11, pp. 1751–1759, 2004. · ·

188. D. A. Horneman, M. Ottens, J. T. F. Keurentjes, and L. A. M. van der Wielen, "Surfactant-aided size-exclusion chromatography for the purification of immunoglobulin G," Journal of Chromatography A, vol. 1157, no. 1-2, pp. 237–245, 2007. · ·

189. N. Suematsu, K. Okamoto, and F. Isohashi, "Simple and unique purification by size-exclusion chromatography for an oligomeric enzyme, rat liver cytosolic acetyl-coenzyme A hydrolase," Journal of Chromatography B, vol. 790, no. 1-2, pp. 239–244, 2003. · ·

190. G. B. Irvine, "High-performance size-exclusion chromatography of peptides," Journal of Biochemical and Biophysical Methods, vol. 56, no. 1–3, pp. 233–242, 2003. · ·

191. J. Strube, R. Gärtner, and M. Schulte, "Process development of product recovery and solvent recycling steps of chromatographic separation processes," Chemical Engineering Journal, vol. 85, no. 2-3, pp. 273–288, 2002. · ·

192. L. S. Pais, J. M. Loureiro, and A. E. Rodrigues, "Modeling strategies for enantiomers separation by SMB chromatography," AIChE Journal, vol. 44, no. 3, pp. 561–569, 1998.

193. C. Migliorini, M. Mazzotti, G. Zenoni, and M. Morbidelli, "Shortcut experimental method for designing chiral SMB separations," AIChE Journal, vol. 48, no. 1, pp. 69–77, 2002. · ·

194. P. S. Gomes, M. Zabkova, M. Zabka, M. Minceva, and A. E. Rodrigues, "Separation of chiral mixtures in real SMB units: the FlexSMB-LSRE," AIChE Journal, vol. 56, pp. 125–142, 2010.

195. O. Ludemann-Hombourger, R. M. Nicoud, and M. Bailly, "The "VARICOL" process: a new multicolumn continuous chromatographic process," Separation Science and Technology, vol. 35, no. 12, pp. 1829–1862, 2000. · ·

196. O. Ludemann-Hombourger, G. Pigorini, R. M. Nicoud, D. S. Ross, and G. Terfloth, "Application of the "VARICOL" process to the separation of the isomers of the SB-553261 racemate," Journal of Chromatography A, vol. 947, no. 1, pp. 59–68, 2002. ·

197. S. R. Perrin, W. Hauck, E. Ndzie et al., "Purification of difluoromethylornithine by global process optimization: coupling of chemistry and chromatography with enantioselective

crystallization," Organic Process Research and Development, vol. 11, no. 5, pp. 817–824, 2007. · ·

198. M. Minceva, P. S. Gomes, V. Meshko, and A. E. Rodrigues, "Simulated moving bed reactor for isomerization and separation of p-xylene," Chemical Engineering Journal, vol. 140, no. 1–3, pp. 305–323, 2008. · ·

199. M. Luká and Z. Pe ina, "A dynamic model of physical processes in chromatographic glucose-fructose separation," Chemical Engineering Science, vol. 46, no. 4, pp. 959–965, 1991.

200. D. C. S. Azevedo and A. E. Rodrigues, "Fructose-glucose separation in a SMB pilot unit: modeling, simulation, design, and operation," AIChE Journal, vol. 47, no. 9, pp. 2042–2051, 2001. · ·

201. H. J. Subramani, K. Hidajat, and A. K. Ray, "Optimization of simulated moving bed and varicol processes for glucose-fructose separation," Chemical Engineering Research and Design, vol. 81, no. 5, pp. 549–567, 2003. · ·

202. S. Imamoglu, "Simulated moving bed chromatography (SMB) for application in bioseparation," Advances in Biochemical Engineering/Biotechnology, vol. 76, pp. 211–231, 2002.

203. J. Andersson and B. Mattiasson, "Simulated moving bed technology with a simplified approach for protein purification: separation of lactoperoxidase and lactoferrin from whey protein concentrate," Journal of Chromatography A, vol. 1107, no. 1-2, pp. 88–95, 2006. · ·

204. P. Li, G. Xiu, and A. E. Rodrigues, "Proteins separation and purification by salt gradient ion-exchange SMB," AIChE Journal, vol. 53, no. 9, pp. 2419–2431, 2007. · ·

205. D. J. Wu, Y. Xie, Z. Ma, and N. H. L. Wang, "Design of simulated moving bed chromatography for amino acid separations," Industrial and Engineering Chemistry Research, vol. 37, no. 10, pp. 4023–4035, 1998.

206. N. Gottschlich and V. Kasche, "Purification of monoclonal antibodies by simulated moving-bed chromatography," Journal of Chromatography A, vol. 765, no. 2, pp. 201–206, 1997. · ·

207. A. Geisser, T. Hendrich, G. Boehm, and B. Stahl, "Separation of lactose from human milk oligosaccharides with simulated

moving bed chromatography," Journal of Chromatography A, vol. 1092, no. 1-2, pp. 17–23, 2005. · ·

208. G. Paredes, M. Mazzotti, J. Stadler, S. Makart, and M. Morbidelli, "SMB operation for three-fraction separations: purification of plasmid DNA," Adsorption, vol. 11, no. 1, pp. 841–845, 2005. · ·

209. Q. Du, P. Wu, and Y. Ito, "Low-speed rotary countercurrent chromatography using a convoluted multilayer helical tube for industrial separation," Analytical Chemistry, vol. 72, no. 14, pp. 3363–3365, 2000. · ·

210. J. Chen, G. X. Ma, and D. Q. Li, "HPCPC separation of proteins using polyethylene glycol-potassium phosphate aqueous two-phase," Preparative Biochemistry and Biotechnology, vol. 29, no. 4, pp. 371–383, 1999.

211. A. E. Herr, J. I. Molho, J. G. Santiago, M. G. Mungal, T. W. Kenny, and M. G. Garguilo, "Electroosmotic capillary flow with nonuniform zeta potential," Analytical Chemistry, vol. 72, no. 5, pp. 1053–1057, 2000. · ·

212. E. Bartow and R. H. Jebens, "Purification of water by electroosmosis," Industrial & Engineering Chemistry, vol. 22, pp. 1020–1022, 1930.

213. C. S. Grant, E. J. Clayfield, and M. J. Matteson, "Surfactant enhanced electro-osmotic separation of iron oxide ultrafines," Colloids and Surfaces, vol. 65, no. 4, pp. 257–272, 1992.

214. P. J. Buijs, A. J. G. van Diemen, and H. N. Stein, "Efficient dewatering of waterworks sludge by electroosmosis," Colloids and Surfaces A, vol. 85, no. 1, pp. 29–36, 1994.

215. S. Al-Asheh, R. Jumah, F. Banat, and K. Al-Zouˌbi, "Direct current electroosmosis dewatering of tomato paste suspension," Food and Bioproducts Processing, vol. 82, no. 3, pp. 193–200, 2004.

216. D. S. Schultz, "Electro-osmotic technology for soil remediation: laboratory results, field trial, and economic modeling," Journal of Hazardous Materials, vol. 55, pp. 81–91, 1997.

217. S. Laursen and J. B. Jensen, "Electroosmosis in filter cakes of activated sludge," Water Research, vol. 27, no. 5, pp. 777–783, 1993. · ·

218. K. Sarangi and A. K. Pattanaik, "A hybrid process for recovering copper from dilute solutions," Separation Science and Technology, vol. 42, no. 1, pp. 89–102, 2007. · ·

219. J. Y. Tian, Z. L. Chen, J. Nan, H. Liang, and G. B. Li, "Integrative membrane coagulation adsorption bioreactor (MCABR) for enhanced organic matter removal in drinking water treatment," Journal of Membrane Science, vol. 352, no. 1-2, pp. 205–212, 2010. · ·

220. Y. C. Wang, M. H. Choi, and J. Han, "Two-dimensional protein separation with advanced sample and buffer isolation using microfluidic valves," Analytical Chemistry, vol. 76, no. 15, pp. 4426–4431, 2004. · ·

221. P. B. Spoor, L. Grabovska, L. Koene, L. J. J. Janssen, and W. R. Ter Veen, "Pilot scale deionisation of a galvanic nickel solution using a hybrid ion-exchange/electrodialysis system," Chemical Engineering Journal, vol. 89, no. 1–3, pp. 193–202, 2002. · ·

222. A. L. Ahmad and S. W. Puasa, "Reactive dyes decolourization from an aqueous solution by combined coagulation/micellar-enhanced ultrafiltration process," Chemical Engineering Journal, vol. 132, no. 1–3, pp. 257–265, 2007. · ·

223. W. Den and C. J. Wang, "Removal of silica from brackish water by electrocoagulation pretreatment to prevent fouling of reverse osmosis membranes," Separation and Purification Technology, vol. 59, no. 3, pp. 318–325, 2008. · ·

224. S. K. Nataraj, K. M. Hosamani, and T. M. Aminabhavi, "Distillery wastewater treatment by the membrane-based nanofiltration and reverse osmosis processes," Water Research, vol. 40, no. 12, pp. 2349–2356, 2006. · ·

225. S. K. Nataraj, K. M. Hosamani, and T. M. Aminabhavi, "Nanofiltration and reverse osmosis thin film composite membrane module for the removal of dye and salts from the simulated mixtures," Desalination, vol. 249, no. 1, pp. 12–17, 2009.

226. S. K. Nataraj, S. Sridhar, I. N. Shaikha, D. S. Reddy, and T. M. Aminabhavi, "Membrane-based microfiltration/electrodialysis hybrid process for the treatment of paper industry wastewater," Separation and Purification Technology, vol. 57, no. 1, pp. 185–192, 2007. · ·

227. AIChE Report on Separation Technology Workshops, Vision 2020: 2000 Separations Roadmap, American Institute of Chemical Engineers, New York, NY, USA, 2000.

Chapter 3

Exploitation of Bacterial Activities in Mineral Industry and Environmental Preservation: An Overview

Ahmed A. S. Seifelnassr[1]
and Abdel-Zaher M. Abouzeid[2]

[1]Department of Mining Engineering, Faculty of Petroleum and Mineral Engineering, Suez Canal University, Suez 62114, Egypt
[2]Department of Mining Engineering, Faculty of Engineering, Cairo University, Giza 12613, Egypt

ABSTRACT

Since the identification and characterization of iron and sulfur oxidizing bacteria in the 1940s, a rapid progress is being made in minerals engineering based on biological activities. Microorganisms can play a beneficial role in all facets of minerals processing, from mining to waste disposal and management. Some of the applications, such as biologically assisted leaching of copper sulfide ores, uranium ores, and biooxidation of refractory sulfide gold ores, are now established on the scale of commercial processes. A variety of other bioleaching opportunities exist for nickel, cobalt, cadmium, and zinc sulfide leaching. Recently, other uses of microorganisms are potentially possible. These include the bioleaching of nonsulfide ores, bioflotation, and bioflocculation of minerals, and bioremediation of toxic chemicals discharged from mineral engineering operations. These activities acquire considerable opportunities for further research and development in these areas. This paper is an attempt to provide a critical summary on the most important efforts in the area of bacterial activities in the mineral and mining industry.

INTRODUCTION

Biotechnology has many potential applications in mining industry including metal leaching, product upgrading, removal of impurities, treatment of acid rock drainage, and other uses for environmental control. Recent interest in the biotechnological processes is the direct application to treat wastes and low-grade ores [1–3]. In this aspect, bacteria catalyze the dissolution of metals from minerals. Therefore, bacterial leaching processes are faster than chemical processes at ambient temperature and atmospheric pressure. So far, only three different types of commercial scale microbiological leaching techniques are practiced for the recovery of copper and uranium from low-grade ores, namely, dump leaching, heap leaching, and insitu leaching. Knowledge about bacterial involvement in these processes has been relatively recent, because the microorganisms responsible for the solubilization of metals from minerals were identified only a few decades ago. Furthermore, heap and dump leaching technologies were introduced in the United States by the Phelps-Dodge Corporation at Bisbee, Arizona, and Tyron, New Mexico, in early 1920s [4], although at that time the processes involved in the leaching and acid drainage production were considered to be solely chemical in nature.

Lately, interest in the biological oxidation of refractory sulfide gold ores has been practiced worldwide [5]. Moreover, microorganisms are used in biobeneficiation which refers to removal of undesirable mineral components from an ore. The interaction with microorganism selectively removes the impurities, and thereby enriches the desired mineral constituent in the solid ore matrix such as, biodesulfurization of coals and biobeneficiation of iron ores. Another potential utilization of microorganisms is that they could be used to flocculate finely divided minerals and/or be used as mineral surface modifiers or flotation collectors.

BIOLEACHING OF SULFIDE ORES

Microorganisms Involved in Leaching Processes

The most important group of bacteria which are involved in sulfide minerals leaching are the acidophilicThiobacilli which belongs to the family Thiobacteriacrae. They have the ability to use the oxidation of inorganic sulfur and its compounds to produce energy for growth. They are, therefore, referred to as chemolithotrophs. They include the autotrophs which derive their carbon for growth solely from carbon dioxide, mixotrophs that can utilize carbon derived from organic compounds, and carbon dioxide, and the heterotrophs whose sole source of carbon is obtained from organic substrates. The majority of the Thiobacillispecies are active between 30 and 35°C. However, moderately thermophilic species have been isolated which grow best at temperature of 45–50°C [6].

In order of importance, the Thiobacilli which are involved in mineral leaching are Acidithiobacillus ferrooxidans, Thiobacillus thiooxidans, Thiobacillus acidophilus, and Thiobacillus oranoporus. Acidithiobacillus ferrooxidans is the most important of the above species [7–9]. This species is able not only to utilize inorganic sulfur compounds but also to oxidize ferrous iron in inorganic substrates. Their differentiation is based upon their capacity to oxidize either elemental sulphur or various sulfide minerals.

Acidithiobacillus ferroxidans is an aerobic, acidophilic autotrophic, Gram-negative, bacterium. It is rod-shaped bacterium and is active above pH 2.0 [10]. Mesophilic strains have an optimum temperature of 35°C for growth. It requires a source of nitrogen, phosphate, and trace amounts of calcium, magnesium, and potassium. Its energy for growth is obtained from the oxidation of ferrous iron, insoluble sulfides, and soluble sulphur compounds. The 9 K nutrient medium was derived for mass production of Acidithiobacillus ferrooxidans cells [11]. There are also some Acidithiobacillus ferrooxidans species which are also acidophilic, autotrophic, rod shaped, mesophilic bacteria, which grow on elemental sulfur, and soluble sulfur compounds, but unable to oxidize ferrous iron or insoluble sulfides.

Thiobacillus acidophilus and Thiobacillus oranoporus are mesophilic, mixotrophic, acidophilus rod-shaped bacteria that oxidize only elemental sulfur for growth. They grow at pH 1.5–5.0 with an optimum value from 2.5 to 3.0. Being unable to oxidize insoluble sulfides, their role in mineral leaching may only be to consume organic compounds excreted by Acidithiobacillus ferrooxidans which are detrimental to the latter organism's growth [12].

In recent years, moderately and extremely thermophilic and acidophilic bacteria which are able to oxidise iron, sulfur and mineral sulfides have been isolated and tested [6, 13–15]. Moderately thermophilic Thiobacilli have been demonstrated to be heterotrophic with optimum temperatures for growth between 45 and 60°C. Strains of thermophilic organisms of a Sulfolobus type grow within a temperature range of 55–85°C. Their role in solubilizing metal is not completely understood. However, the usefulness of thermophilic microorganisms can also be extended to bioremediation activities.

Mechanisms of Bioleaching

Some doubt still surrounds the exact role of bacteria in the oxidation of sulfide minerals because of the inability to discretely separate reactions which are solely promoted by bacteria from those which are simply chemical. The concept of direct and indirect modes of bacterial leaching of metal sulfides (MS) was introduced few decades ago [10].

In the direct mode of bacterial leaching mechanism, the sulfide is oxidized to metal sulfate:

$$MS + 2O_2 \xrightarrow{\text{bacteria}} MSO_4 \tag{1}$$

where M is a bivalent metal. The heavy metal sulfides are generally insoluble in aqueous acid leach media while their sulfates are soluble. In some cases, the oxidation product is insoluble as, for example, in the case of lead sulfide leaching. This fact can be utilized for selective leaching [16] to separate soluble zinc, copper, and cadmium from insoluble lead. In the direct mode of bacterial oxidation, bacteria must remain close to the surface of the solid substrate.

In the indirect mode, ferric ion produced from bacterial oxidation of pyrite, which is always associated with sulfide minerals, is the oxidant. The sequence of reactions is as follows:

$$2FeS_2 + 7.5O_2 + H_2O \xrightarrow{\text{bacteria}} Fe_2(SO_4)_3 + H_2SO_4 \tag{2}$$

$$MS + Fe_2(SO_4)_3 \xrightarrow{\text{chemical}} MSO_4 + S^o + 2FeSO_4 \tag{3}$$

$$2FeSO_4 + H_2SO_4 + 0.5O_2 \xrightarrow{\text{bacteria}} Fe_2(SO_4)_3 + H_2O \tag{4}$$

$$S^o + 1.5O_2 + H_2O \xrightarrow{\text{bacteria}} H_2SO_4 \tag{5}$$

In the absence of bacteria, elemental sulfur deposited on the surface of the particles may grow in proportion so as to create a thick enough layer to inhibit the progress of the leaching process. The sulfuric acid produced may further react with the oxide contents (MO) of the ore, thus contributing to the metal dissolution process:

$$MO + H_2SO_4 \xrightarrow{\text{chemical}} MSO_4 + H_2O \tag{6}$$

An example of indirect bacterial leaching activity is the oxidation of chalcopyrite, $CuFeS_2$, in the presence of pyrite. In this process, the copper mineral is leached in the presence of bacteria in the following manner:

$$CuFeS_2 + 2Fe_2(SO_4)_3 \xrightarrow{\text{chemical}} CuSO_4 + 5FeSO_4 + 2S^o \tag{7}$$

Again the reaction by-products ferrous iron and sulfur are oxidized by bacteria to ferric iron and sulfuric acid following reactions (4) and (5).

The growth of Acidithiobacillus ferrooxidans is measured by cell count of the supernatants of the suspensions, whereas the extent of bacterial attachment/adsorption to minerals during leaching was estimated from cell protein concentration of the solid and liquid phases. Probable mechanism of attachment and detachment of bacteria was also discussed [17, 18]. Recently, a two-step mechanism for bioleaching was proposed [19, 20]. It involves chemical ferric reaction with the mineral to produce ferrous salt, and then bacterial oxidation of ferrous iron to ferric completes a closed loop of reactions.

Developments in Bioleaching of Sulfide Ores

In 1947, Acidithiobacillus ferrooxidans, the main microorganism responsible for metal sulphide oxidation was first isolated and characterized [21]. It was found that this bacterium could oxidize the sulfide part of the mineral to sulfuric acid and the ferrous ion to ferric ion. This oxidation ability can be demonstrated in the oxidation of pyrite, which is almost always found with the sulfide minerals equation (2).

The bioleaching of pyrite will be discussed later when referring to coal desulfurization in Section 4 entitled biobeneficiation. Numerous systematic studies [22–25] have subsequently revealed that Acidithiobacillus ferrooxidans, under acidic leaching conditions, can attack most sulfide minerals, producing water-soluble metal sulphates. The optimum leaching conditions can be summarized as follows: growth media (nutrients) [26, 27], temperature 35°C [28, 29], pH 2.3 [30], Eh below 500 mV in order to avoid jarosite precipitation [31,32], high specific surface area of solids [33, 34], and prior adaptation of bacteria to specific substrate [23, 35,36]. Large scale heap and dump leaching operations were built so as to provide the best growth conditions for the microorganisms in order to harvest their beneficial effects in dissolution of metal from mining wastes [37,38].

Most investigations concerning the bioleaching of copper from low-grade ores have been conducted in the laboratory using small columns or simulated in large scale tests. The influence of variations in the mineralogical composition and textural features of copper ores as well as process variables have been examined [39, 40]. Chalcopyrite is leached in the presence of bacteria in (7).

Again, the above reaction by-products, ferrous iron and sulfur, are oxidized by bacteria to ferric iron and sulfuric acid.

The oxidation mechanisms for chalcocite (Cu_2S) can be expressed by the following equation:

$$Cu_2S + 0.5O_2 + H_2SO_4 \xrightarrow{\text{bacteria}} CuS + CuSO_4 + H_2O$$

$$(8)$$

$$CuS + 2O_2 \xrightarrow{\text{bacteria}} CuSO_4$$

(9)

In addition, extensive studies were conducted with thermophilic microorganism in the temperature range of 45–85°C [41–43]. The advantage of using thermophilic organisms in the leaching of sulfide minerals is that, at higher temperature, the reaction kinetics is expected to increase. A new genus of thermophilic spore-forming bacteria, sulfobacillus, was reported [43].

Due to the refractory nature of the chalcopyrite, the utilization of high temperatures and thermophilic bacteria has been investigated. It is reported that typical copper extraction yields obtained by mesophilic bacteria are about 30%, whereas copper extraction yields of more than 98% can be obtained in shorter periods by thermophilic bioleaching [44, 45]. A study concerning bioleaching of chalcopyrite showed that the bioleaching of chalcopyrite is controlled by the oxidation-reduction potential, temperature, pH, and the activity of the thermophile used [46, 47].

Recently, a comparative study [48] on the bioleaching of chalcopyrite concentrates using mesophilic and moderately thermophilic bacteria indicated that the moderately thermophilic bacteria have higher ability for copper dissolution. These results show that copper dissolution from the chalcopyrite concentrate reached 87.52% with the moderately thermophilic bacteria while it was 34.55% with mesophilic culture after 25 days.

The applicability of bacterial leaching technique to the recovery of uranium from low-grade ores has been investigated [49, 50]. In bacterial leaching of uranium ores, the tetravalent uranium is oxidized to its hexavalent state, which is soluble, by ferric sulfate:

$$UO_2 + Fe_2(SO_4) + 2H_2SO_4$$

$$\xrightarrow{\text{chemical}} H_4[UO_2(SO_4)_3] + 2FeSO_4$$

(10)

The role of bacteria is to reoxidize ferrous iron to the ferric state. Ferric sulfate is obtained by metabolic oxidation of pyrite, which is always present in the uranium ores. Bioleaching of copper and uranium ores by heap leaching resulted in substantial saving in the production costs.

Bioleaching of zinc sulfide concentrates using bacteria has been investigated [16, 51]. The following reaction is proposed:

$$ZnS + 2O_2 \xrightarrow{\text{bacteria}} ZnSO_4 \tag{11}$$

The maximum rate of zinc extraction, under optimum conditions, was around 640 mg/dm³ h in terms of specific surface area, particle size, and pulp density of the solid substrate. Selective extraction of zinc, copper, and cadmium from below the cut-off grade (complex) lead sulfide concentrates is illustrated in the flow diagram in Figure 1. The method is especially applicable to ores with very fine crystalline intergrowth of lead, zinc, cadmium, and copper sulfides where quantitative recovery from individual mineral fractions is not possible by physical separation techniques. The leach residue in PbS concentrate leaching in this operation is a high-grade lead concentrate, which consists of unreacted PbS and insoluble $PbSO_4$. The recovery step my involve precipitation of iron by increasing the pH to 3.5 using lime. Copper and cadmium are obtained by cementation and zinc hydroxide is precipitated by increasing pH value to 7.5 using magnesia. Zinc hydroxide can be converted to zinc by acidification and electrowinning [16, 23, 35, 52].

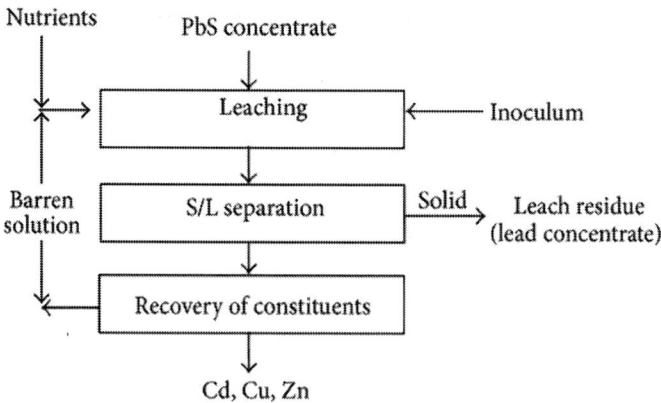

Figure 1: Schematic representation of a selective bacterial leaching process of a complex lead sulfide concentrate (S/L signifies solid-liquid separation) [16].

Almost complete extraction of pentlandite, using the microorganisms Acidithiobacillus ferrooxidans, can be expressed by

$$(Ni, Fe)_9S_8 + 17.6250_2 + 3.25H_2SO_4$$

$$\xrightarrow{bacteria} 4.5NiSO_4 + 2.25Fe_2(SO_4)_3 + 3.25H_2O$$

(12)

It was possible to dissolve cobalt and nickel at a high rate from the sulfide minerals and to produce Co^{+2} and Ni^{2+} ion concentrations as high as $30\,g/dm^3$ and $71\,g/dm^3$, respectively. Selective extraction of arsenic from a complex, finely disseminated stannic, auriferous, zinc-copper ore has been described [16, 53]. A basic flow diagram of this process is shown in Figure 2. In this process the arsenic content of the ore is solubilized by bacteria and, after solid-liquid separation, it is precipitated by addition of lime to raise the pH to about 3.0. The dissolved copper is recovered by cementation with scrap iron and the solution is recycled. From the solid residue of bacterial leaching, the unreacted copper ore is removed by flotation, yielding copper sulfide concentrate and a tin enriched residue. The process in Figure 2 can also be applied for leaching of gold-arsenic sulfides from finely disseminated metal in the sulfide matrix. In this case, the precious metals remain in the residue from bacterial leaching. The residue is neutralized by addition of lime and treated with cyanide solution to dissolve gold and silver. From the leach solution, arsenic is precipitated and discarded. The bacterial leaching in these studies can be considered as a preoxidation step which exposes the precious metals for subsequent cyanidation or thiourea leaching [5].

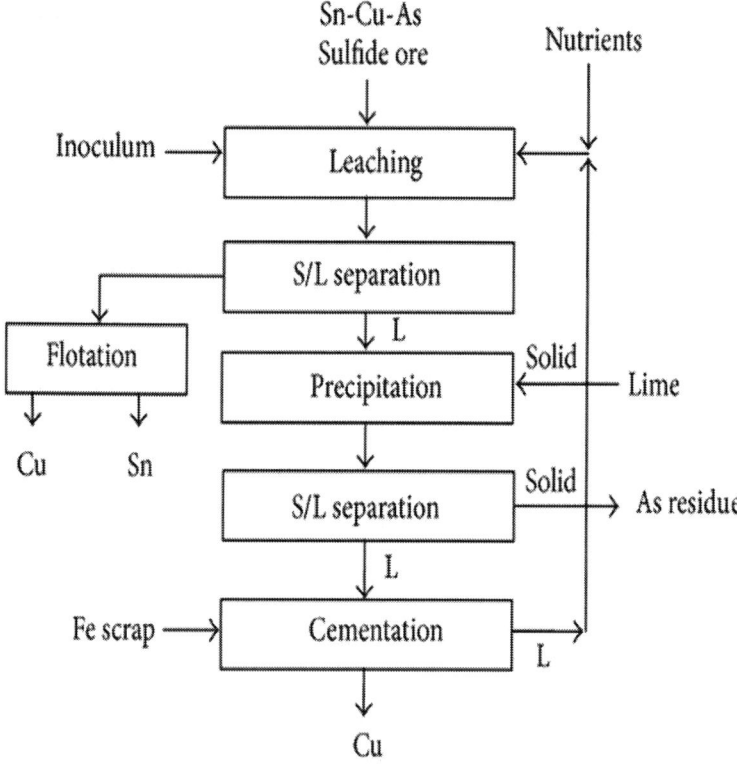

Figure 2: Bacterial leaching process for difficult-to-dress Sn-As-Cu ores [16].

Lately, bacterial leaching methods gained further impetus with the introduction of biopreoxidation processes for the liberation of precious metals from sulfide-bearing minerals [54–56]. If gold occurs in a finely disseminated form within the sulfide ore matrix, the economic viability of conventional gold extraction processes by cyanide leaching becomes less than marginal. Extensive research work has been carried out for the treatment of the complex gold-bearing sulfide ores. It is reported that pyrite oxidation by bioleaching improved gold recovery. This promising improvement was proportional to the degree of oxidation (Figure 3). For example, with 84% oxidation of pyrite, gold recovery in solution reached 81%.

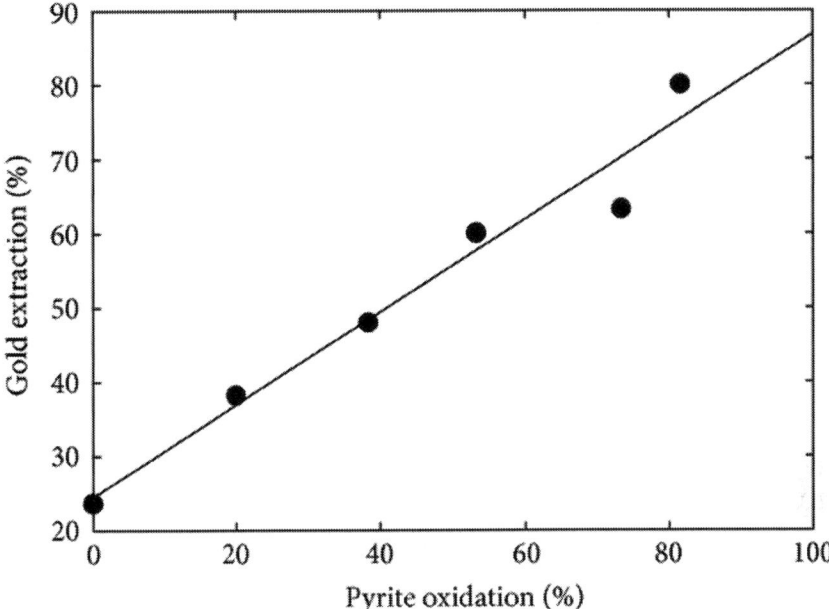

Figure 3: Effects of biological pyrite oxidation on gold recovery from Porgera concentrate [57].

Bioleaching research has demonstrated that microorganisms can tolerate exceptionally high metal ion concentration (120 g/L zinc, 72 g/L nickel, 30 g/L cobalt, 55 g/L copper, and 12 g/L U_3O_8) and high hydrogen ion concentration (acid media of pH range 1–5) during leaching of sulfide minerals [43].

Genetic manipulations of leaching organisms were initiated in the eighties [16, 58]. The purpose of these studies was to develop specific metal extraction using microorganisms capable of a high rate of metabolic conversion and resistant to toxic elements. Biosorption technology using free or immobilized alive or dead cells or their derivatives in films, aggregates, or pellets was illustrated. It was found that biosorption technology is especially applicable to the removal of toxic metal contamination from large volume of industrial waste streams containing trace amounts of heavy metals and radionuclide [59].

BIOPROCESSING OF REFRACTORY GOLD ORES

Bacterial leaching processes will be significant in the treatment of difficult-to-process refractory ores [60–63]. The gold in refractory ores is encapsulated as fine particles in the crystal structure of sulfide matrix such as pyrite (FeS_2) and arsenopyrite (FeAsS). This makes the efficiency of the cyanidation process very low since the cyanide solution cannot penetrate the sulfide-bearing gold crystals and dissolve gold particles, even after fine grinding. Therefore, an oxidative pretreatment is necessary to decrease the refractory properties of the ore. Roasting is sometimes used, but it is highly energy consuming and involves a costly off-gas neutralization system to prevent atmospheric pollution [64, 65]. Both pressure oxidation and oxidation by nitric acid require high temperature and/or corrosion-resistant materials which are costly items. Hence, biological pretreatment becomes an interesting alternative route. This route leads to environmental protection and low-cost processes [64]. Research and developments in this direction have been stimulated by the buoyant price of the precious metal and also by the fact that conventional methods of extraction are not able to produce a sufficiently high recovery of the contained value [66]. It has been demonstrated by both laboratory and pilot test work that such process is feasible [67]. The biological pretreatment of refractory gold ores is based on the ability of some microorganisms such as Acidithiobacillus ferrooxidans and Thiobacillus thiooxidans to oxidize and dissolve the bearing-gold sulfide minerals, thus liberating the entrapped gold particles, thereby rendering it amenable to the cyanidation process [57, 64]. Advancements in this area have been made on industrial scale to improve the rate of oxidation and to reduce cyanide consumptions in downstream gold recovery [68]. The results indicated a direct relationship between the degree of sulphide mineral oxidation and percent gold recovery. Complete oxidation of sulphides is not necessary to achieve significant enhancement of gold recovery. On the basis of the sulphide entity, high gold recoveries can be obtained with as low as 50% oxidation of the total sulphides.

BIOBENEFICIATION

Biobeneficiation refers to removal of undesirable mineral components from an ore through interactions with microorganisms which bring about their selective removal by a bioleaching process. Compared to bioleaching of sulfide minerals by Thiobacilli, bioleaching of nonsulfide minerals has received little attention in the past. For example, desulfurization of coal, bioleaching of aluminum from aluminosilicates, removal of alumina and silica from iron ores, and so forth have been extensively studied. These interactions lead to enriching these desired mineral constituents in the solid ore matrix mediated by a number of surface chemical and physiochemical phenomena. The mediation roles include alteration of the surface chemistry of minerals, generation of metabolic products which cause chemical dissolution, selective dissolution of mineral phases in an ore matrix, and sorption, accumulation, and precipitation of ions and compounds on solid surfaces.

In order to minimize the potential deleterious impact of increased amounts of sulfur dioxide emission due to coal burning, the sulfur content of coal must be reduced. The biodesulfurization of coal presents a potentially attractive alternative to chemical and physical methods [60, 69, 71]. In the biodesulfurization process, the pyrite content of coal will be oxidized to water-soluble ferric sulfate and sulfuric acid according to (2). The dissolved ferric sulfate is removed from the coal in the dewatering step. The coal is then washed and dried prior to combustion.

Experimental investigations indicated that bacteria and fungi could be effectively used to remove iron and silica from clays, sands, and bauxite ores [15, 72, 73]. Successful commercialization of bauxite biobeneficiation was proposed [74]. Biological removal of calcium and iron from a low-grade bauxite ore was discussed with respect to Bacillus polymyxa. Growth conditions and probable mechanisms in the biological removal of calcium and iron from the bauxite ore were outlined by Anand et al. [75]. From the reported results, changes in the pH of the leach medium correlated well with the calcium dissolution. The presence of bacteria lowers the pH and hence facilitates calcium dissolution.

Iron ores generally contain alumina, silica, sulfur, and phosphorous as the main gangue minerals. These impurities have adverse effects on reducibility of iron oxides, coke rate consumption, and blast furnace operation and productivity for steel making. Various studies have examined the use of the heterotrophic bacteria and fungi for removal of alumina and silica from iron ores for improving the iron content of the concerned ore. The iron ore beneficiation was carried out by secondary metabolites produced by these heterotrophic microorganisms [76, 77]. It has been reported that in situ leaching of an iron ore with fungal strains such as Aspergillus fumigatus, Penicillium citrinum, and Aspergillus flavus resulted in 7%, 6%, and 17% removal of alumina, and 8%, 4%, and 16% removal of silica, respectively. Bacillus polymyxa, Bacillus sphaericus, and Pseudomonas putida ensured silica removal percentage of 10.6%, 5.3%, and 20%, respectively.Aspergillus flavus and Pseudomonas putida were most efficient among all the bacterial and fungal strains used, ensuring an increase in iron content of about 3% at the end of 10 days leaching [78].

Ronini [79] reported that heterotrophic organisms can be used to leach out the alumina and silica from the slimes generated by Tata Iron and Steel Company in India. He investigated the feasibility of Bacillus to leach the slimes and increase its iron content. At pH 7, leaching for 5 days, at inoculums size of 20%, Ronini obtained an optimum recovery of 79% of the iron content in the slimes.

BIOSURFACE MODIFICATION

Adhesion of microorganisms to mineral surfaces is known to alter the hydrophobicity of minerals. It has been demonstrated that Acidithiobacillus ferrooxidans is suitable for the rapid treatment of sulfide ores where leaching is not the desired outcome. Surface treatment of sulfide minerals with bacterial solution is shown to influence their superficial chemical properties, thus altering their response in processes such as froth flotation and/or selective flocculation. This technique is being evaluated as a method of enhancing the physical separation of pyrite from coal in fine coal flotation circuits and is suggested as an alternative method to the total leaching of pyrite from coal [80, 81]. In this technique the coal pulp is conditioned with Acidithiobacillus ferrooxidans bacteria for about 30 minutes and thus renders pyrite surface to be hydrophilic. This, in turn, enhances the selective flotation of coal from pyrite. Table 1 shows typical results of a study concerning bacterial leaching versus bacterial conditioning followed by flotation of minus 28 mesh coal containing 2.88% pyritic sulfur [69].

Table 1: Flotation, bacteria leaching, and combinations of bacterial conditioning and flotation of −28 mesh coal containing 2.88% pyritic sulfur [69]*

Process	Coal product specifications				
	Coal yield, %	Pyritic sulfur, %	Pyritic sulfur removal, %	**Ash, %**	**Calorific value, kcal/kg**
Regular conditioning and one-stage flotation	73.74	1.29	66.76	22.5	—
Bacterial leaching (10 days leaching)	100	1.42	56.6	30.47	5260
Bacterial conditioning (4 hours) and one stage flotation	78.0	0.825	77.63	18.5	6361
Bacterial conditioning (4 hours) and 3-stage flotation	34.36	0.68	91.78	12.03	—

*pH = 2.0 for flotation conditioning and for bacterial leaching, and pH = 9 for all flotation stages.

In a study concerning the effect of bacterial conditioning of sphalerite and galena, it was found that the floatability of galena decreased markedly (Figure 4) due to oxidation of sulfur to insoluble lead sulfate species on the surface [82, 83]. In the case of sphalerite (Figure 4) such effects were not observed since the zinc sulfate formed is soluble. The reported results have significant implications to the selective flotation of lead-zinc sulfides.

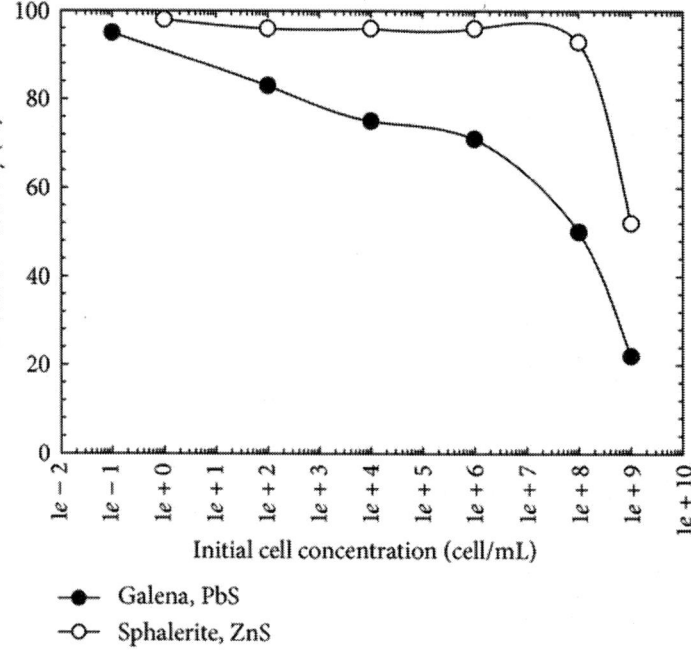

Figure 4: Effect of initial cell concentration during bacterial conditioning on the floatability of galena and sphalerite [82].

Recently, a copper concentrate assaying 22.23% Cu was obtained through bacterial conditioning followed by flotation, whereas a copper concentrate assaying 18.20% Cu was obtained in conventional flotation [84]. This means that the copper grade of the flotation concentrate, subjected to bacterial conditioning, is higher by 22% than the concentrate obtained by conventional flotation without bacterial conditioning. Acidithiobacillus ferrooxidans can affect mineral surfaces by direct (intimate) contact or indirect (no intimate) contact mechanisms. In both cases, the bacteria eliminate the occurrence of oxidized sulfur which (the sulfur) has hydrophobic properties and induces higher floatability to minerals so that hydrophobicity of pyrite is decreased [85].

According to the mechanisms explained above, bacteria are more effective on the pyrite surface than on the chalcopyrite surface. This is because at low pH values, the oxidation of pyrite is more pronounced than that of chalcopyrite. In addition, Acidithiobacillus ferrooxidans increases the oxidation rate of pyrite gradually. Under these conditions, the formation of jarosite layer takes place at lower pH values. Once jarosite is formed, it precipitates on mineral surfaces and decreases the effectiveness of reagent/mineral surface interaction in flotation resulting in pyrite depression. From the above discussion, it could be conclude that Acidithiobacillus ferrooxidans appears to play a dual role, promoting flotation under certain conditions while enhancing depression of minerals under some other conditions. Promotion of floatability of sulphide minerals in the presence of this type of bacteria could be understood in the light of elemental sulphur formation on mineral surfaces through biooxidation. Bacterial interaction for prolonged periods of time leads to reoxidation of the sulphur to sulphoxy compounds and ultimately to sulphate. Gradual build-up of such oxidized layers on mineral surfaces would impede flotation.

BACTERIA ACTIVITIES IN FLOTATION AND FLOCCULATION

There is high evidence that microorganisms could be used to flocculate finely divided minerals and/or other solids suspensions [86, 87]. It was found that the bacterium, Mycobacterium phlei, has a demonstrated potential to be used for the flotation of hematite, Figure 5. The decrease in flotation recovery at high bacteria concentration (>20 ppm) was due to the formation of hematite aggregates too large to be levitated by air bubbles [88]. This same type of bacterium proved to be successful in flocculating a variety of finely divided minerals such as hematite (Figure 6), phosphate slimes (Figure 7), and coal (Figure 8) [86, 87, 89]. Figure 6shows that the concentration of bacteria cells affects the extent of flocculation. This type of bacterium,Mycobacterium Phlei, possesses highly negative features on highly hydrophobic surfaces. It was suggested that these properties arise in large part from its fatty acid surface [88]. Because of these characteristics, the organism which is readily adsorbed onto the hydrophilic surface of the mineral may have a negative, neutral, or low positive charge. It also adheres to many hydrophobic minerals due to the created adhesion (attraction) energy of hydrophobic interactions [90].

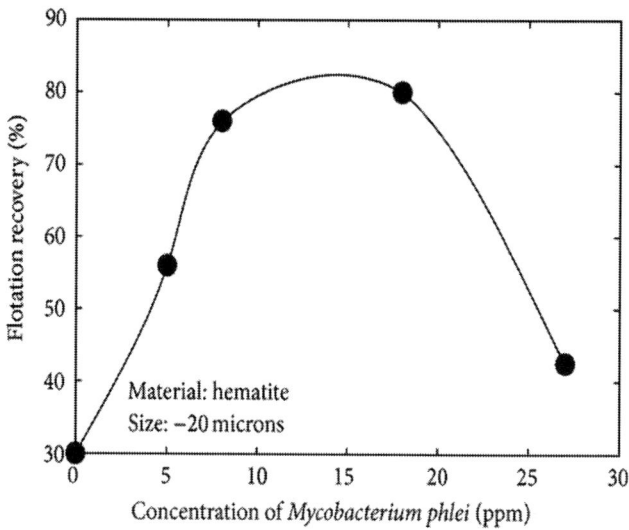

Figure 5: Hallimond tube flotation recovery of hematite as a function of Mycobacterium phlei concentration (operating conditions: pH = 5; 1 gram of hematite; 10 min flotation) [88].

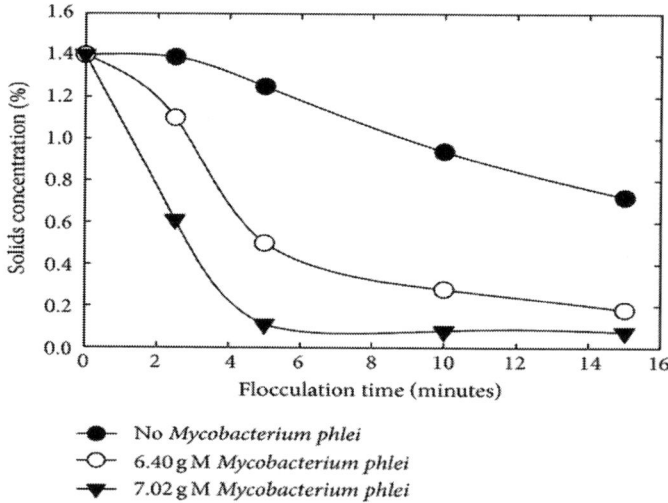

Figure 6: Flocculation of hematite slimes with and without Mycobacterium phlei as a function of time. Samples were collected at 4 cm from the bottom surface of a 1000 mL graduated cylinder [88].

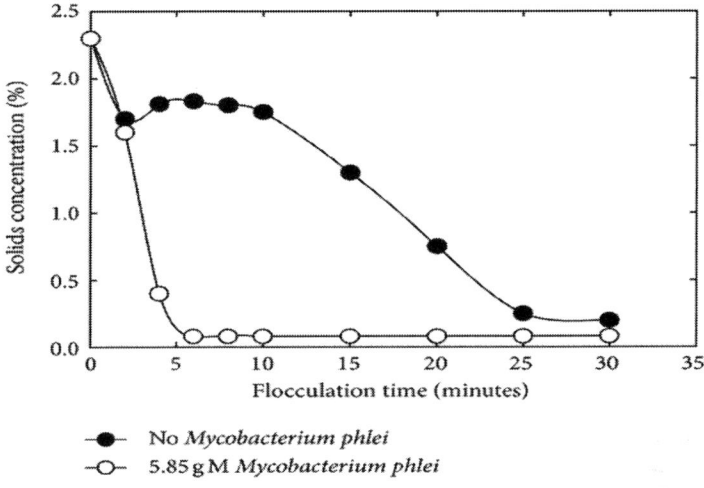

Figure 7: Flocculation of a 1.4% suspension of Four Corners (Florida) phosphate slime with the addition of two different concentrations of Mycobacterium phlei [88].

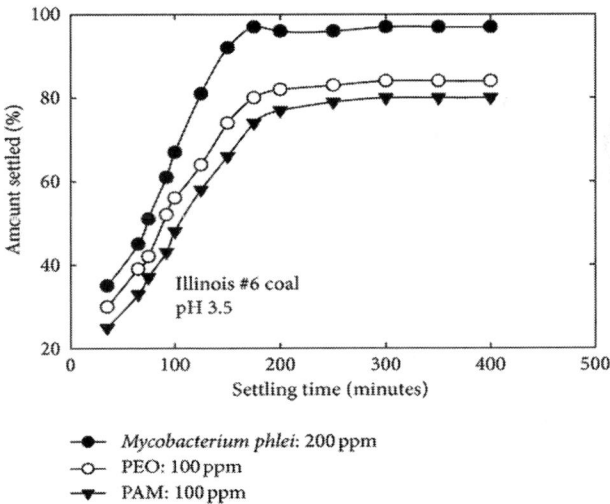

Figure 8: Amount of coal settled as a function of time in the presence of different flocculants: Mycobacterium phlei, Polyacrylamide flocculent (PAM), and Polyethylene Oxide flocculent (PEO) [89].

Interaction between Paenibacillus polymyxa with minerals such as hematite, corundum, calcite, kaolinite, and quartz resulted in significant surface—chemical changes. Quartz and kaolinite were rendered more hydrophobic, while hematite, calcite, and corundum became more hydrophilic after biotreatment. Through biotreatment of the above minerals, it was possible to selectively separate silica and alumina from iron minerals either by flotation or selective flocculation [91].

Utilization of microorganisms and associated extracellular metabolic products in selective flotation and flocculation has been recently reported [91–93]. Patra and Natarajan [94] showed that different protein fractions derived from Paenibacillus polymyxa exhibited varying surface adsorption capacity towards minerals such as quartz, pyrite, chalcopyrite, galena, and sphalerite. Proper use of fractionated protein groups rendered pyrite and chalcopyrite hydrophilic, while sphalerite, galena and quartz exhibited enhanced surface hydrophobicity after bio-treatment. Similarly, prior protein treatment resulted in selective flocculation of pyrite and chalcopyrite together, while galena, sphalerite, and quartz were effectively dispersed. These studies demonstrated that bacterial proteins could effectively replace the conventional amine and xanthate types of collectors which are toxic and expensive.

Due to the adherence of bacteria to mineral surfaces, some strains can be used to modify mineral surfaces to aid selective recovery of valuable minerals in flotation or flocculation processes. Some bacteria can selectively depress the flotation of one mineral compared to another. The depression can either result from bacteria oxidizing or otherwise modifying the surface of the mineral to render it less floatable or prevent the subsequent adsorption of a flotation collector. Due to the adherence of bacteria to mineral surfaces, some strains can be used to modify mineral surfaces to aid selective recovery of valuable minerals in flotation or flocculation processes. Some bacteria can selectively depress the flotation of one mineral compared to others. The depression can either result from bacteria oxidizing or otherwise modifying the surface of the mineral to render it less floatable or from bacteria adhering to the mineral preventing the subsequent adsorption of a flotation collector.

Adhesion of Bacillus subtilis and Mycobacterium phlei onto dolomite and apatite was studied by sorption measurements and scanning electron microscopy [95]. It was found that both Bacillus subtilis andMycobacterium phlei adhere onto dolomite surface more readily than onto apatite surface at acidic and near neutral pH values. At more basic pH values Bacillus subtilis adheres more readily onto the mineral surface and remains a better depressant for dolomite than for apatite. However, Mycobacterium phlei, at basic pH values, adsorbs more onto apatite than onto dolomite acting as a weaker depressant for dolomite and a stronger depressant for apatite compared with Bacillus subtilis. The differences in adsorption characteristics were attributed to differences in surface properties of the two bacteria species and of the two minerals. The net result of the study indicated that, while both bacteria function as depressants in anionic collector flotation of dolomitic phosphate ores, Bacillus subtilis functions as the stronger depressant, especially for dolomite [95].

In a more recent investigation, Sarvamangala and Natarajan [70] showed that the microorganism Bacillus subtilis and the extracellular protein have been utilized for the separation of hematite from the other oxide minerals. It is evident from the obtained results that the presence of bacterial cells and cell-free extract promoted the flocculation and settling of hematite whereas in the case of quartz, corundum, and calcite the interaction with bacterial cells and cell-free extract favored more dispersion of the minerals. Flotation behavior of hematite-quartz and calcite-corundum systems was studied before and after interaction with bacterial cell-free extract and bacterial cells. The obtained results, Table 2 [70], indicate that interaction with Bacillus subtilisconfers surface hydrophobicity on quartz, calcite, and corundum, while similar biotreatment renders hematite more hydrophilic. Relative hydrophobicity or hydrophilicity of mineral-grown bacterial cells depends on the ratio of proteins and polysaccharides present on the cell walls. Bacterial cell population and their interaction period with minerals, as well as mineral surface coverage through bacterial adhesion control the mineral surface hydrophobicity regarding flotation and/or flocculation. These studies open a wide venue for possible developments of biotechnological applications for environmentally safe mineral beneficiation operations. However, more detailed investigations need to be carried out to make a clear insight into the control of bacterial cell wall composition. Also, the mechanisms of bacteria-mineral surfaces interactions should be clearly highlighted.

Table 2: Flotation recovery of minerals treated individually, in presence and in absence of collector without cells or extract and after interaction with mineral-grown cells and cell-free extract [70]

Mineral (-105+75 microns)	Flotation recovery, percent			
	Without cells or cell-free extract		With cells or cell-free extract	
	Without collector	With collector	After interaction with cells (1 h)	After interaction with cells-free extract (1 h)
Quartz	14.6	97.8	91.5	90.1
Calcite	12.5	95.0	74	50.0
Corundum	11.0	96.0	73.2	30.0
Hematite	11.0	95.0	4.8	14

In 2011, Reyes-Bozo et al. [96] studied the effect of biosolids (obtained from waste water treatment plant, Chile) on hydrophobic properties of sulfide ores on a laboratory scale. The principal components of biosolids are humic substances, mainly humic acid, and phosphorus compounds. The interaction between the mineral surface and the functional groups found in biosolids, as a collector, for copper sulfide ores, was investigated through zeta potential measurements, FT-IR analysis, and film flotation tests. The results showed that biosolids change the hydrophobicity of the sulfide minerals by adsorbing onto the surface. Biosolids show greater affinity for pyrite while commercial humic acid shows similar behavior to industrial collectors. Therefore, both biosolids and humic acids can change the hydrophobic properties of sulfide ores and can be used as collectors in froth flotation processes. Thus, the use of biosolids is feasible in a preliminary flotation stage for removing pyrite or in the rougher stage of froth flotation to separate important sulfide minerals from the gangue.

BIOREMEDIATION

In addition to being useful in the mineral beneficiation area, recent developments in biotechnology have given promises that biotechnology may also provide means for bioremediation of environmental problems generated in the mineral, metallurgical, and chemical industries. For example, in the flotation of wastes, an investigation concerning the biodegradation of thiol collectors by the bacterium Pseudomonas fluorescents was reported [88]. In this study it was indicated that a residual xanthate concentration of 0.12 mg/L in the wastewater from a lead concentrator was completely destroyed in five minutes after treatment with a bacterial suspension. Butyl xanthate destruction by Pseudomonas fluorescents has also been investigated, and typical results are shown in Figure 9. In this figure, it is clear that the concentration of butyl xanthate was decreased by about 20% of its original concentration in 40 minutes in the presence of bacteria [88].

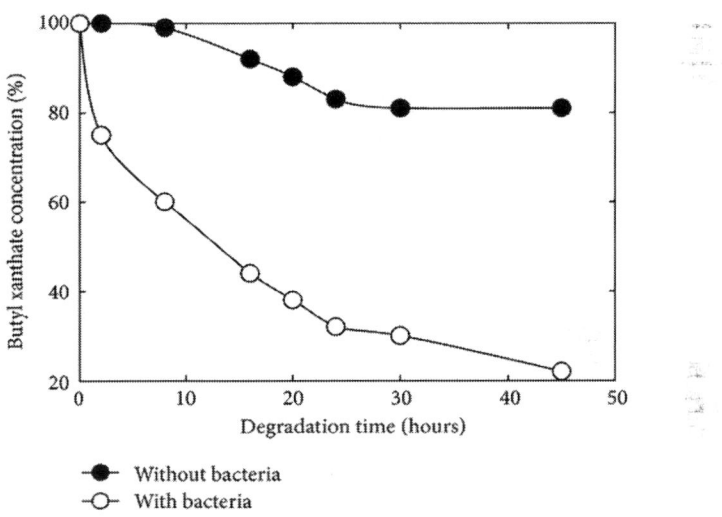

Figure 9: Influence of Pseudomonas fluorescens on the degradation of butyl xanthate [88].

Concerning polluted soil bioremediation, there is an excellent review, in which sources of soil pollution, bioremediation strategies, and the direction of further research have been highlighted [97]. It is known that, under specified conditions, certain microorganisms or enzymes derived from microorganisms are able to break down cyanides, and hence, there is a potential for using these organisms in bioremediation cyanide wastes discharged from precious metal hydrometallurgical plants [98]. Noel et al. [24] cultivated bacterial strains from solids previously exposed to cyanide solution which tolerate 300 ppm sodium cyanide under anaerobic conditions. In the selected soil samples, these bacteria reduced the level of cyanide from approximately 300 ppm to essentially zero in about 50 days under anaerobic conditions. Typical results of these experiments are shown in Figure 10. In this study, various nutrient media were investigated and the maximal growth of bacteria was established at Medium A, which was composed of 1.0 g/L K_2HPO_4, 0.2 g/L $MgSO_4 \cdot 7H_2O$, 2 g/L $FeSO_4$, 2 g/L $MnCl_{12} \cdot 4H_2O$, and 0.001 g/L $Na_2MoO_4 \cdot 2H_2O$.

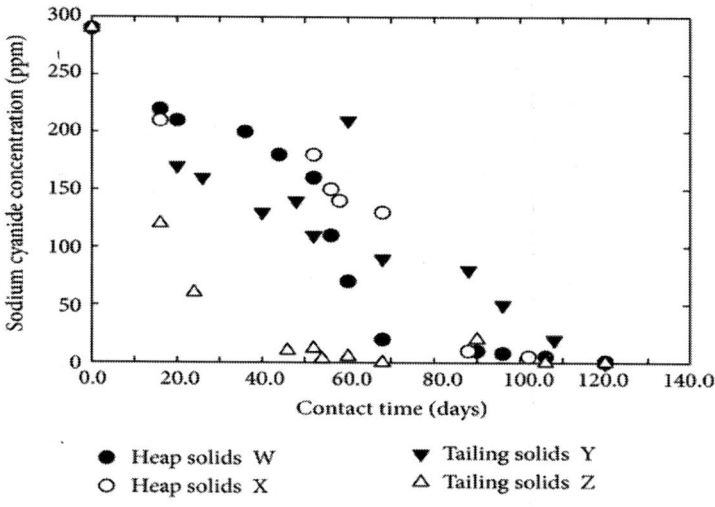

Figure 10: Variation of sodium cyanide concentration as a function of inoculum solids and contact time in the presence of nutrient Medium A at locations W, X, Y, and Z [24].

Maniatis et al. [99] demonstrated that biological destruction of cyanide in mining water was effective in the laboratory and in the field. In this study, the cyanide was put in a complex form with selenium by aerobic reactor which was run continuously for six days to remove cyanide and then run one more time through a series of anaerobic reactors to remove selenium. The aerobic reactor removed 95% of the cyanide content in the first 24 hours with another 3% removal over the next five days. No nutrient addition was required.

Biosorption laboratory research activities are expected to reach industrial application for the detoxication of industrial waste water

[100, 101]. Considerable interest exists in the application of biosorption to the removal of a number of environmental contaminants including toxic heavy metals such as chromium [102], selenium [103], and cadmium, as well as radionuclides such as uranium [104].

Chaalal et al. [105] reported the use of thermophilic bacteria (belongs to Bacillus family) for the removal of lead compounds contaminating the drinking water. These bacteria were isolated and used in a reactor coupled with a membrane system. The bacteria, the stirrer, and the membrane housed in the reactor were arranged in a distinctive way to form the novel biostabilization process proposed in this research. They claimed that the proposed technique could be used at low cost and with great confidence in purifying drinking water. The system was found to be adequate for remediating drinking water having lead concentration up to 40 ppm. At the end of the operation, the lead concentration reaches the level allowed by the world health organization regulations.

Bioremediation of waters contaminated with crude oil and toxic heavy metals was also achieved by the process of microbial dissimilatory sulfate reduction and biosorption [106].

SUMMARY

The present paper highlights a number of new possibilities for industrial application of biotechnological principles for the extraction of metal values from inorganic resources. The present industrial interest in bioleaching methods is motivated by the fact that these processes can produce metal values from low-grade resources for approximately one-third to one-half of the cost of the conventional smelting techniques without polluting the environment. Furthermore, selective implementation of living systems can offer opportunities for reduced labor, increased productivity, and technological advances. In fact, bacteria technologies have been applied on a commercial scale for the recovery of copper and uranium from low-grade ores and industrial wastes.

Bacterial activities have been recently introduced in the mineral processing technology. They have been used in mineral surface modification, flocculation, and collectors in flotation. In these areas, parameter optimization and process control are required for efficient application. In addition, higher levels for scaling up the operations must be encouraged in all biomineral processing systems for industrial purpose. Also, detailed understanding of the biosurfactant role in modifying the mineral surfaces and exploring bioflotation applications is currently lacking.

Bacterial leaching is a new development in hydrometallurgy of metal sulphides. Greater understanding of the basic mechanisms of bacterial mineral leaching is still required which hinders the controlled utilization of this technology. Published information relating to the oxidation of various minerals differs greatly and sometimes contradicts each other, with respect to the acting mechanisms, the rate, and amount of oxidation took place under specific conditions. This would indicate that considerable information is still necessary to determine the most efficient types of bacteria, proper operating conditions, manner in which they should be cultivated, and the mechanisms with which they attach to the minerals to be leached.

Microbiological leaching is influenced by a number of parameters, and it functions best if carried out at optimum leaching conditions. Intensive aeration is required to assure maximum oxygen mass transfer into the leach solution. The smallest particle size of the solid sulphide substrate will assure the highest rate and yield of metal extraction. The Eh must be kept below 500 mV in order to avoid jarosite type and basic ferric hydroxide precipitation on the surface of the solid substrate. In all leaching techniques, wherever possible, a cyclic leaching process should be applied.

It seems that new trends are evolving as the knowledge of bacterial activity relative to industrial applications is furthered. Biohydrometallurgical applications are multidisciplinary in nature, and should be dealt with at optimum conditions for optimal effects. Bioremediated processes are of high importance, especially in copper, uranium, and precious metals industries. However, their applicability in the remediation of contaminated soils and industrial aqueous effluents is hampered by lack of sufficient information.

REFERENCES

1. Q. P. Granger, "Bacterial leaching of minerals," Colliery Guardian Redhill, vol. 232, no. 6, pp. 212–214, 1984.

2. F. D. Pooley, "The role of biohydrometallurgy in mineral processing," in Innovations in Mineral and Coal Processing, S. Atak, G. Onal, and M. S. Celik, Eds., p. 435, Balkema, Rotterdam, The Netherlands, 1998.

3. A. S. S. Seifelnassr and A. Z. M. Abouzeid, "New trends in mineral processing: exploitation of bacterial activities," The Journal of Mineral Processing, vol. 3, no. 4, p. 17, 2000.

4. A. W. Hudson and G. D. Vanasdale, "Heap leaching at Bisbee, Arizona," Transactions of the Society of Mining, vol. 64, p. 137, 1923.

5. A. Bruynesteyn and R. P. Hack, "The biotank leach process for the treatment of refractory gold/silver concentrates," in Microbiological Effects on Metallurgical Processes, J. A. Clum and L. A. Haas, Eds., pp. 121–128, Tms-AIME, New York, NY, USA, 1985.

6. J. Murpby, E. Ristenberg, D. Marek, R. Moble, B. Beck, and D. Skidmore, "Microbial dessulphurization of coal by Thermophilic bacteria," in Microbiological Effects on Metallurgical Processes, J. A. Clum and L. A. Haas, Eds., pp. 99–110, TMS, 1985.

7. J. E. Moss and J. E. Anderson, "The effect of environment on bacterial leaching rates," Proceedings of the Australasian Institute of Mining and Metallurgy, vol. 225, p. 15, 1968.

8. M. Makintosh, "Nitrogen fixation by T. ferrooxidans," Journal of General Microbiology, vol. 70, p. 66, 1971.

9. A. E. Torma, "The role of Thiobacillus ferrooxidans in hydrometallurgical processes," Advances in Biochemical Engineering, vol. 6, pp. 1–37, 1977.

10. M. P. Silverman, "Mechanism of bacterial pyrite oxidation," Journal of Bacteriology, vol. 94, no. 4, pp. 1046–1051, 1967.

11. M. P. Silverman and D. G. Lundgren, "Studies on the chemoautotrophic iron bacterium ferroobacillus ferrooxidans an improved medium and harvesting procedure for securing high cell yields," Journal of Bacteriology, vol. 77, pp. 642–647, 1959.

12. F. D. Pooley, "Mineral leaching with bacteria," in Environmental Biotechnology, F. F. Christopher and D. A. John, Eds., pp. 114–134, Ellis Horwood, John Wiley and Sons, New York, NY, USA, 1987.

13. C. L. Brierley and J. A. Brierley, "A chemoautotrophic and thermophilic microorganism isolated from an acid hot spring," Canadian Journal of Microbiology, vol. 19, no. 2, pp. 183–188, 1973.

14. G. Millonig, M. De Rosa, A. Gambacorta, and J. D. Bu'lock, "Ultrastructure of an extremely thermophilic acidophilic micro organism," Journal of General Microbiology, vol. 86, no. 1, pp. 165–173, 1975.

15. V. I. Groudeva, S. N. Grouder, and M. I. markov, "A comparison between Thermophilic bacterial with respect to their ability to leach sulfide minerals," in Fundamental and Applied Biohydrometallurgy, R. W. Lawrence, R. M. Brauion, and H. G. Ebener, Eds., p. 484, Elsevier, 1986.

16. A. E. Torma, "Biohydrometallurgy as an emerging technology," in Proceedings of the Biotechnology and Bioengineering Symposium No. 16, p. 49, 1986.

17. M. L. Free, T. Oolman, S. Nagpal, and D. A. Bahlstrom, "Bioleaching of sulfide ores—distinguishing between indirect and direct mechanisms," in Mineral Bioprocessing, R. W. Smith and M. A. Misra, Eds., p. 485, TMS, 1991.

18. Y. R. K. Mirajkar, K. A. Natarajan, and P. Somasundaran, "Growth and attachment of Thiobacillus ferrooxidans during sulfide mineral leaching," International Journal of Mineral Processing, vol. 50, no. 3, pp. 203–210, 1997.

19. G. S. Hansford, "Studies on the mechanisms and kinetics of bioleaching," Fizykochemiczne Problemy Mrtalugil, vol. 32, pp. 281–291, 1998.

20. D. Mishra and Y. Rhee, "Current research trends of microbiological leaching for metal recovery from industrial wastes," in Current Research, Technology, Education Topics in Applied Microbiology and Microbial Biotechnology, A. Mendez-Vilas, Ed., FORMATEX, 2010.

21. A. R. Colmer and M. E. Hinkle, "The role of microorganisms in acid mine drainage: a preliminary report," Science, vol. 106, no. 2751, pp. 253–256, 1947.

22. W. R. Ruzzel and P. C. Trussel, "Isolation and properities of an iron oxidizing Thiobacillus," Journal of Bacteriology, vol. 85, p. 595, 1963.

23. K. A. Natarajan and I. Iwasaki, "Microbe/mineral interaction in leaching of complex sulfides," inMicrobiological Effects on Metallurgical Processes, S. A. Clum and L. A. Hass, Eds., p. 113, Tms-AIME, New York, NY, USA, 1985.

24. D. M. Noel, M. C. Fuerstenau, and J. L. Hendrix, "Degradation of cyanide utilizing facultative anaerobic bacteria," in Mineral Bioprocessing, R. W. Smith and M. Misra, Eds., pp. 355–366, TMS, 1991.

25. W. E. Ruzzel, "Bacterial leaching of metallic sulfides," Canadian Institute of Mining, vol. 55, p. 190, 1962.

26. N. Lazaroff, "Sulfate requirement for iron oxidation to enhance gold and silver recovery from pyritc ores and concentrates," CIM Bulletin, vol. 85, p. 78, 1963.

27. A. H. Tuovimen and D. P. Kelly, "Studies on the growth of Thiobacillus ferrooxidans," Archives of Microbiology, vol. 88, p. 285, 1973.

28. I. J. Corrans, B. Harris, and B. J. Ralph, "Bacterial leaching: an introduction to its application and theory and a study on its mechanisms of operation," Journal of the South African Institute of Mining and Metallurgy, vol. 72, p. 221, 1972.

29. A. Pinches, "Bacterial leaching of an arsenic-bearing sulfide concentrate," in Leaching and Reduction in Hydromrtallurgy, A. R. Burkin, Ed., p. 28, IMM, London, UK, 1975.

30. H. Sakaguchi and M. Silver, "Microbiological leaching of a chalcopyrite concentrate by Thiobacillus ferrooxidans," Biotechnology and Bioengineering, vol. 18, no. 8, pp. 1091–1101, 1976.

31. A. E. Torma, C. C. Walden, and R. M. Branion, "Microbiological leaching of a zinc sulfide concentrate,"Biotechnology and Bioengineering, vol. 12, no. 4, pp. 501–517, 1970.

32.   C. L. Brierley, "Bacterial leaching," CRC Critical Reviews in Microbiology, vol. 6, no. 3, pp. 207–206, 1978.

33.   R. L. Braun and R. G. Mallon, "Combined leach-circulation calculation for predicting in-situ copper leaching of primary sulfide ore," Transactions of the Society of Mining Engineers AIME, vol. 258, no. 2, pp. 103–110, 1975.

34.   P. R. Norris, L. Parrott, and R. M. Marsh, "Moderately Thermophilic mineral-oxidizing bacteria," inProceedings of the Biotechnology and Bioengineering Symposium No. 16, H. L. Ehrlich and D. S. Holmes, Eds., pp. 253–363, John Wiley and Sons, 1986.

35.   H. Kandemnir, "Fate of sulfide Sulfur bacterial oxidation of sulfide minerals," in Microbiological Effects on Metallurgical Processes, J. A. Clum and L. A. Haas, Eds., p. 51, TMS, 1985.

36.   M. Elzeky and Y. A. Attia, "Effect of bacterial adaptation on kinetics and mechanisms of bioleaching ferrous sulfides," Chemical Engineering Journal and the Biochemical Engineering Journal, vol. 56, no. 2, pp. B115–B124, 1995.

37.   E. Peters, "Thermodynamic and kinetic factors in the leaching in sulfide minerals from ore deposits and dumps," SME Short Course in Bio Extractive Mining, SME/AIME, 1970.

38.   A. Bruynesteyn and J. R. Copper, "Leaching of Canadian ore in test deposits," in Proceedings of the Solution Mining Symposium, F. F. Aplon and W. A. Mchinezy, Eds., p. 268, 1974.

39.   A. A. S. Seifelnassr, Bacterial aided percolation leaching of copper sulfide ores [Ph.D. thesis], University of Wales, Cardiff, UK, 1988.

40.   A. A. S. Seifelnassr and F. D. Pooley, "Biologically assisted ferric ion leaching of refractory copper sulfide ore," in Proceedings of the V111 International Mineral Processing Symposium, Antalya, Turkey, October 2000.

41.   J. A. Brierley and C. L. Brierley, "Microbial leaching of copper at ambient and elevated temperatures," inMetallurgical Applications of Bacterial Leaching and Related Microbiological Phenomenena, L. E. Murr, A. E. Torma, and J. A. Brierley, Eds., pp. 477–489, Academic Press, London, UK, 1978.

42. L. E. Murr, A. E. Torma, and J. A. Brieley, Metallurgical Applications of Bacterial Leaching and Related Microbiological Phenomena, Academic Press, New York, NY, USA, 1978.

43. H. M. Tsuchiya, "Microbial leaching of Cu-Ni sulfide concentrate," in Metallurgical Application of Bacterial Leaching and Related Microbiological Phenonena, L. E. Murr, A. E. Torma, and J. A. Brierley, Eds., pp. 365–372, Academic Press, London, UK, 1978.

44. M. Gericke, A. Pinches, and J. V. Van Rooyen, "Bioleaching of a chalcopyrite concentrate using an extremely thermophilic culture," International Journal of Mineral Processing, vol. 62, no. 1–4, pp. 243–255, 2001.

45. A. Sissing and S. T. L. Harrison, "Thermophilic mineral bioleaching performance: a compromise between maximizing mineral loading and maximizing microbial growth and activity," Journal of The South African Institute of Mining and Metallurgy, vol. 103, no. 2, pp. 139–142, 2003.

46. J. Vilcáez, K. Suto, and C. Inoue, "Bioleaching of chalcopyrite with thermophiles: temperature-pH-ORP dependence," International Journal of Mineral Processing, vol. 88, no. 1-2, pp. 37–44, 2008.

47. J.-L. Xia, Y. Yang, H. He et al., "Investigation of the sulfur speciation during chalcopyrite leaching by moderate thermophile Sulfobacillus thermosulfidooxidans," International Journal of Mineral Processing, vol. 94, no. 1-2, pp. 52–57, 2010.

48. A. Behrad Vakylabad, "A comparison of bioleaching ability of mesophilic and moderately thermophilic culture on copper bioleaching from flotation concentrate and smelter dust," International Journal of Mineral Processing, vol. 101, no. 1–4, pp. 94–99, 2011.

49. W. A. Gow and G. M. Ritcey, "Treatment of canadian uranium ores," Canadian Mining and Metallurgical Bulletin, vol. 62, no. 692, pp. 1330–1339, 1969.

50. R. Guay, A. E. Torma, and M. Silver, "Ferrous ion oxidation and uranium solubilization from a lowgrade ore by "Thiobacillus ferrooxidans"," Annales de Microbiologie, vol. 126, no. 2, pp. 209–219, 1975.

51. A. E. Torma, C. C. Walden, D. W. Duncan, and M. R. Brauion, "Effect of carbon dioxide and particle surface area on the micro

biological leaching of a zinc sulfide concenytates," Biotechnology and Bioengineering, vol. 14, p. 777, 1992.

52.   A. E. Torma and K. N. Subramanian, "Selective bacterial leaching of a lead sulphide concentrate,"International Journal of Mineral Processing, vol. 1, no. 2, pp. 125–134, 1974.

53.   Y. Attia, L. Tchfield, and L. Vaaler, "Application of bio-technology in the recovery of gold," inMicrobiological Effects on Metallurgical Processes, J. A. Clum and L. A. Haas, Eds., pp. 11–20, Tms-AIME, New York, NY, USA, 1985.

54.   E. Livesey, P. Norman, and R. Livesey, "Gold recovery from arsenopyrite/pyrite ore by bacterial leaching and cyanidation," in Recent Progress in Biohydrometallurgy, pp. 627–641, Associozione Mineraria Sarda, Iglesias, Italy, 1983.

55.   E. Livesey, "Bacterial leaching of gold, uranium, pyrite-bearing-compacted mine tailing slimes," inFundamental and Applied Biouhydro Metallurgy, R. W. Lawrnce, R. M. Braniou, and H. G. Ebmer, Eds., pp. 89–97, Elsevier, 1986.

56.   H. L. Ehrlich, "Bacterial leaching of silver from a silver containing mixed Sulfide ore by a continuous process," in Fundamental and Applied Biohydrometallurgy, R. W. Lawrence, R. M. Braniou, and H. G. Ebmer, Eds., pp. 77–88, Elsevier, 1986.

57.   R. W. Lawrence and A. Bruynesteyn, "Biological pre-oxidation to enhance gold and silver recovery from refractory pyritic ores and concentrates," CIM Bulletin, vol. 76, no. 857, pp. 107–110, 1983.

58.   D. S. Holmes and K. A. Debus, "Opportunities for biological metal recovery," in Mineral Bioprocessing, R. W. Smith and M. Misra, Eds., pp. 57–80, Tms-AIME, 1991.

59.   C. C. Towskey, I. S. Ross, and A. S. Atkins, "Biorecovery of metallic residues from various industrial effluents using filamentous Fungi," in Fundamental and Applied Biohydromrtallurgy, R. W. Lawrence, R. M. R. Branion, and H. G. Ebner, Eds., pp. 279–290, Elsevier, 1986.

60.   A. E. Torma, "Mineral bioprocessing," in BIOMIN, 93, pp. 1–10, Australian Mineral Foundation, Glenside, South Australia, 1993.

61.   S. N. Groder, I. I. Spasova, and I. M. Ivauov, "Microbial leaching of a gold-bearing pyrite Concentrate," in Changing Scopes in Mineral Processing, M. Kemal, V. Arslan, A. Askar, and M.

Canbazolgu, Eds., pp. 583–586, Balkema, Rotterdam, The Netherlands, 1996.

62. A. Ozkan, S. Aydogan, and U. Akdermir, "Bacterial leaching as a pre-treatment step for gold recovery from refractory ores," in Proceedings of the Physicochemical problems of Mineral Processing, vol. 32, pp. 173–182, Wroclaw, Poland, 1998.

63. Z. Sadowski, T. Farbiszewska, and J. Farbiszewka-Bajar, "The role of microorganisms in pretreatment of gold-bearing ores," in Proceedings of the Physicochemical Problems of mineral Processing 35th Symposium, pp. 151–165, Wroclaw, Poland, 1998.

64. S. Ubaldini, F. Vegliό, L. Toro, and C. Abbruzzese, "Biooxidation of arsenopyrite to improve gold cyanidation: study of some parameters and comparison with grinding," International Journal of Mineral Processing, vol. 52, no. 1, pp. 65–80, 1997.

65. D. Karamanev, A. Margaritis, and N. Chong, "The application of ore immobilization to the bioleaching of refractory gold concentrate," International Journal of Mineral Processing, vol. 62, no. 1–4, pp. 231–241, 2001.

66. B. V. Mihaylov and J. L. Hendrix, "Biooxidation of a sulfide gold ore in columns," in Mineral Bioprocessing, R. W. Smith and M. Misra, Eds., p. 163, TMS-AIME, 1991.

67. B. A. Paponetti, S. Ubaldini, C. Abbruzzese, and L. Tora, "Biometallurgy for the recovery of gold from arsenopyrite Ores," in Mineral Bioprocessing, R. W. Smith and M. Misra, Eds., p. 179, TMS, 1991.

68. P. Miller and A. Brown, "Bacterial oxidation of refractory gold concentrates.," in Advances in Gold Ore Processing, M. A. Adams, Ed., Elsevier, 2005.

69. M. Z. Dogan and M. S. Cleik, "Latest developments in coal desulphurization by flotation and microbial beneficiation," in Proceedings of the 3rd Mining, Petroleum, and Metallurgical Conference, vol. 1, pp. 2–4, Faculty of Engineering, Cairo University, February 1992.

70. H. Sarvamangala and K. A. Natarajan, "Microbially induced flotation of alumina, silica/calcite from haematite," International Journal of Mineral Processing, vol. 99, no. 1–4, pp. 70–77, 2011.

71. T. Farbiszewska, "Intensity of the bacterial leaching process from mining brown coal waste," Physico-Chemical Problems of Mineral Processing, vol. 22, pp. 145–159, 1990.

72. G. I. Karavviko, Z. A. Avakyan, L. V. Ogurtsova, and O. F. Safanova, "Microbiological processing of bauxite," in Proceedings of International Symposium on Biohydrometallurgy, J. Salley, R. G. L. McGready, and P. L. Wichlacz, Eds., pp. 93–102, Canmet, Ottawa, Canada, 1989.

73. L. V. Ogurtsova, G. I. Karavaiko, Z. A. Avakyan, and A. A. Korenevsii, "Activity of various microorganisms in extracting elements from bauxite," Microbiology, vol. 58, pp. 774–780, 1990.

74. S. S. Vasan, J. M. Modak, and K. A. Natarajan, "Some recent advances in the bioprocessing of bauxite,"International Journal of Mineral Processing, vol. 62, no. 1–4, pp. 173–186, 2001.

75. P. Anand, J. M. Modak, and K. A. Natarajan, "Biobeneficiation of bauxite using Bacillus polymyxa: calcium and iron removal," International Journal of Mineral Processing, vol. 48, no. 1-2, pp. 51–60, 1996.

76. C. Cameselle, M. T. Ricart, M. J. Núñez, and J. M. Lema, "Iron removal from kaolin. Comparison between "in situ" and "two-stage" bioleaching processes," Hydrometallurgy, vol. 68, no. 1–3, pp. 97–105, 2003.

77. H. L. Ehrlich, "Past, present and future of biohydrometallurgy," Hydrometallurgy, vol. 59, no. 2-3, pp. 127–134, 2001.

78. S. Shitarashmi, Biomineral processing: a suitable approach [M.S. thesis], National Institute of Technology, Rourkela, India, 2009.

79. N. Ronini, Feasibility study on the microbial separation of iron ore slime [M.S. thesis], National Institute of Technology, Rourkela, India, 2011.

80. G. F. Andrews, P. R. Dugan, and C. J. Stevens, "Combining physical and bacterial treatment for removing pyritic sulfur from coal," in Processing and Utilization of High Sulphur Coal IV, P. R. Dugan, D. R. Quigley, and Y. A. Attia, Eds., p. 515, Elsevier, 1991.

81. Y. A. Attia, M. Elzekey, F. Bavariam, and L. S. Fan, "Cleaning, and desulphurization of high sulfur coal by selective flocculation and

bioleaching in draft tube fluidized bed reactor," in Proceedings of the 3rd Mining, Petroleum, Metallurgy Conference, vol. 1, pp. 2–4, Faculty of Engineering, Cairo University, February 1992.

82. M. K. Yelloji, K. A. Natarajan, and P. Somasundran, "Effect of bacterial conditioning of sphalerite and galena with Thiobacillus ferrooxidans on their floatability," in Mineral Bioprocessing, R. W. Smith and M. Misra, Eds., pp. 105–120, TMS, 1991.

83. K. Hanumantha Rao, A. Javadi, T. Karlkvist, A. Patra, A. Vilinska, and I. V. Chernyshova, "Revisiting sulphide mineral (Bio) processing: a few priorities and directions," in Proceedings of the 15th Balkan Mineral Processing Congress, Sozopol, Bulgaria, June 2013.

84. A. Ekrem Yüce, H. Mustafa Tarkan, and M. Zeki Doğan, "Effect of bacterial conditioning and the flotation of copper ore and concentrate," African Journal of Biotechnology, vol. 5, no. 5, pp. 448–452, 2006.

85. L. C. Bryner, R. B. Walker, and R. Palmer, "Some factors influencing the biological oxidation of sulfide minerals," Transactions of AIME, vol. 238, pp. 56–62, 1967.

86. M. Misra, S. Chen, and R. W. Smith, "Kerogen aggregation using a hydrophobic bacterium," in Mineral Bioprocessing, R. W. Smith and M. Misra, Eds., p. 133, TMS-AIME, 1991.

87. M. Misra, R. W. Smith, and J. Dubel, "Bioflocculation of finely divided minerals," in Mineral Bioprocessing, R. W. Smith and M. Misra, Eds., p. 91, TMS-AIME, 1991.

88. R. W. Smith and M. Misra, "Mineral bioprocessing—an overview," in Mineral Bioprocessing, W. R. Smith and M. Misra, Eds., pp. 3–26, TMS, 1991.

89. M. A. Raichur, M. Misra, and R. W. Smith, "The Potential for selective flocculation of coal from pyrite using a Hydrophic bacterium," in Mineral Processing, Recent Advances and Future Trends, S. P. Mehrotra and R. Shekhar, Eds., pp. 686–693, Allied, New Delhi, India, 1995.

90. D. A. Elgillani, Class Notes in Surface Chemistry, Cairo University, Faculty of Engineering, Department of Mining, Petroleum, and Metallurgical Engineering, Giza, Egypt, 2008.

91. K. A. Natarajan and N. Deo, "Role of bacterial interaction and bioreagents in iron ore flotation,"International Journal of Mineral Processing, vol. 62, no. 1–4, pp. 143–157, 2001.

92. D. Santhiya, S. Subramanian, K. A. Natarajan, H. Hanumantha Rao, and K. S. E. Forssberg, "Bio-modulation of galena and sphalerite surfaces using Thiobacillus thiooxidans," International Journal of Mineral Processing, vol. 62, no. 1–4, pp. 121–141, 2001.

93. M. N. Chandraprabha, K. A. Natarajan, and P. Somasundaran, "Selective separation of pyrite from chalcopyrite and arsenopyrite by biomodulation using Acidithiobacillus ferrooxidans," International Journal of Mineral Processing, vol. 75, no. 1-2, pp. 113–122, 2005.

94. P. Patra and K. A. Natarajan, "Role of mineral specific bacterial proteins in selective flocculation and flotation," International Journal of Mineral Processing, vol. 88, no. 1-2, pp. 53–58, 2008.

95. X. Zheng, P. J. Arps, and R. W. Smith, "Adhesion of two bacteria onto dolomite and apatite: their effect on dolomite depression in anianic flotation," International Journal of Mineral Processing, vol. 62, no. 1–4, pp. 159–172, 2001.

96. L. Reyes-Bozo, R. Herrera-Urbina, M. Escudey et al., "Role of biosolids on hydrophobic properties of sulfide ores," International Journal of Mineral Processing, vol. 100, no. 3-4, pp. 124–129, 2011.

97. S. Pal, A. K. Patra, S. K. Reza, W. Wildi, and J. Pote, "Use of bio-resources for bioremediation of soil pollution," Natural Resources, vol. 1, pp. 110–125, 2010.

98. S. Copaescu, G. fodor, G. Bota, L. Popa, and A. Pescaru, "Possibilities of treatment of residual waters containing cyanide and its recovery in a cyanidation plant from regia autonoma a cupului deva," inChanging Scopes in Mineral Processing, M. Kemal, V. Arslan, A. Akar, and M. Canbozoglu, Eds., pp. 591–598, Balkema, Rotterdam, The Netherlands, 1996.

99. T. Maniatis, B. Wahlquist, and T. Pickett, "Biological cyanide destruction in mineral processing waters," in Proceedings of the SME Annual Meeting, pp. 879–880, Denver, February 2004.

100. J. A. Brierley, C. L. Brierley, and G. M. Goyalc, "AMT-BIOCLAM,

a new waste water treatment and metal recovery technology," in Fundamental and Applied Biohydrometallurgy, R. W. Lawrence, R. M. R. Branion, and H. G. Ebner, Eds., pp. 291–304, Elsevier, 1986.

101. T. Jeffers, C. R. Ferguson, and P. G. Bennett, "Biosorption of metal contaminants from acidic mine waters," in International Mineral Bioprocessing, R. W. Smith and M. Misra, Eds., p. 289, TMS, 1991.

102. W. A. Apel and C. E. Turick, "Bio-remediation of hexavalent chromium by bacterial reduction," inMineral Bio-Processing, R. Smith and M. Misra, Eds., p. 376, TMS-AIME, 1991.

103. J. M. Barnes, E. B. McNew, J. K. Polman, J. H. McCune, and A. E. Torma, "Selenate reduction by pseudomonas stutzeri," in Mineral Bioprocessing, R. W. Smith and M. Misra, Eds., p. 367, TMS-AIME, 1991.

104. M. L. Apel, J. M. Barnes, and A. E. Torma, "Biosorption kinetics of metal removal from uranium mill tailing effluents," in Bio-Processing, R. Smith and M. Misra, Eds., p. 339, TMS, 1991.

105. O. Chaalal, A. Y. Zekri, and R. Islam, "Uptake of heavy metals by microorganisms: an experimental approach," Energy Sources, vol. 27, no. 1-2, pp. 87–100, 2005.

106. V. I. Groudeva, S. N. Groudev, and A. S. Doycheva, "Bioremediation of waters contaminated with crude oil and toxic heavy metals," International Journal of Mineral Processing, vol. 62, no. 1–4, pp. 293–299, 2001.

Fly Ash and Composted Biosolids as a Source of Fe for Hybrid Poplar: A Greenhouse Study

Kevin Lombard[1], Mick O'Neill[1], April Ulery[2], John Mexal[2], Blake Onken[3], Sue Forster-Cox[4], and Ted Sammis[2]

[1]Agricultural Science Center, New Mexico State University, P.O. Box 1018, Farmington, NM 87499, USA

[2]Department of Plant and Environmental Sciences, New Mexico State University, P.O. Box 30003, MSC 3Q, Las Cruces, NM 88003-8003, USA

[3]Lindsay Corporation, 2222 North 111th Street, Omaha, NE 68164, USA

[4]Department of Health Science, New Mexico State University, P.O. Box 30001, MSC 3HLS, Las Cruces, NM 88003-0136, USA

ABSTRACT

Soils of northwest New Mexico have an elevated pH and $CaCO_3$ content that reduces Fe solubility, causes chlorosis, and reduces crop yields. Could biosolids and fly ash, enriched with Fe, provide safe alternatives to expensive Fe EDDHA (sodium ferric ethylenediamine di-(o-hydroxyphenyl-acetate)) fertilizers applied toPopulus hybrid plots? Hybrid OP-367 was cultivated on a Doak sandy loam soil amended with composted biosolids or fly ash at three agricultural rates. Fly ash and Fe EDDHA treatments received urea ammonium nitrate (UAN), biosolids, enriched with N, did not. Both amendments improved soil and plant Fe. Heavy metals were below EPA regulations, but high B levels were noted in leaves of trees treated at the highest fly ash rate. pH increased in fly ash soil while salinity increased in biosolids-treated soil. Chlorosis rankings improved in poplars amended with both byproducts, although composted biosolids offered the most potential at improving Fe/tree growth cheaply without the need for synthetic inputs.

INTRODUCTION

The New Mexico State University Agricultural Science Center at Farmington, San Juan County, has been exploring short rotation hybrid poplar trees for fiber and timber production, biofuel, and phytoremediation purposes. Adaptability trials involving numerous Populus crosses have produced a range of responses. Of these, Fe deficiency chlorosis (interveinal yellowing of juvenile leaves) has been observed because soil pH can exceed 8 with moderate to high $CaCO_3$ levels. Under these conditions, soil iron is mostly in the form of well-crystallized iron oxides (e.g., hematite and goethite) and almost insoluble and unavailable to plants [1]. On our research plots, chelated iron fertilizer in the form of Fe EDDHA is applied to alleviate chlorosis symptoms. Considering that 5 kg Fe EDDHA material—enough to cover approximately 1 ha season^{-1}—costs approximately $200, fertilizing large-scale plantations may be cost prohibitive.

On the other hand, fly ash, a byproduct from coal combustion, can provide plant-available Fe and other micronutrients [2–7]. Fly ash exits the combustion chamber with the flue gas and is captured by electrostatic precipitators, wet scrubbers, or other mechanical/chemical trap [8]. Particle sizes range from 0.01 to 100 µm allowing a large amount of surface area to mass [9]. Nearly 3.9 million Mg of coal combustion products (ash + flue gas desulfurization products) are produced in San Juan County each year by two coal fired generating plants, and both power plants are actively seeking recycling options (Salisbury, 2003, personal communication).

Biosolids (dewatered sewage sludge) also increase levels of plant-available Fe on calcareous soils [6, 10, 11] and are a source of other plant-essential elements including N and P [12, 13]. Iron enhancement in biosolids results from multiple factors at the wastewater treatment facility. When washed into treatment plants through storm runoff, iron oxides can be reduced and reprecipitated as weakly crystalline plant-available iron phosphates [14]. Salts of $FeCl_3$ or $FeCl_2$ used to capture phosphorus from the waste stream during the treatment process also increase the iron phosphate content of biosolids [14, 15]. The city of Albuquerque, 290 km southeast of Farmington, produces 142 Mg of biosolids per day and is a regional leader in seeking land-use disposal and marketing options of processed, composted biosolids [16].

Environmental consequences for both byproducts also have been documented. Fly ash can contain elevated levels of heavy metals, increase boron to toxic levels, can act as liming agents because of their high Ca/Mg content, and can increase soil salinity [17, 18]. Biosolids also have the potential to increase salinity, heavy metals, and persistent organic pollutants such as antibiotics, and personal care products that enter the waste stream [19, 20]. If an environmentally responsible use can be established, recycling of these byproducts to agricultural lands may present an attractive disposal alternative because of the large land area devoted to crop production within a relatively short distance from the power plant or wastewater treatment facilities in the Farmington area. The objectives of this study were to pilot test the application of fly ash and composted biosolids at three rates to a high pH soil from Northwest, NM. Specific objectives were to

- determine if Fe nutrition of soil and the growth of hybrid poplar clone OP-367 could be improved by amending soil with each amendment,
- examine potential environmental issues, including heavy metal contamination, salinity, and pH changes in the soil, caused by each amendment.

MATERIALS AND METHODS

Soil and Treatments

2004 Study Soil and Treatments

A Doak sandy loam (fine-loamy, mixed, mesic Typic Haplargid) [21] was collected from the top 20–25 cm of the plow layer from an agricultural field located at the New Mexico State University Agricultural Science Center, Farmington (lat. 36° 41' 0" N; long. 108° 18' 36" W; elevation 1,700 m). Soil was sieved through 6-mm × 6-mm mesh to remove clods then transported to the NMSU Fabian Garcia horticulture farm greenhouse complex (Las Cruces, NM). Prior to container filling, a fiberglass mesh screen was used to line standard 7.5 L nursery containers to prevent soil loss through drainage holes. Each container was filled to a dry weight of 9 kg. Once filled, the surface area at the top of each container was 366 cm^2.

Fly ash was collected from the APS Four Corners Power Plant (Farmington, NM). Fly ash at the power plant is stored in a lined ash impoundment area adjacent to plant. Composted biosolids were collected from the City of Albuquerque, NM Pilot Composting Facility. The biosolids were a 1 : 3 ratio of dewatered sewage sludge mixed with chipped yard waste that were composted in large windrows at 57°C for six weeks. The composting process reduces pathogen concentrations to comply with USEPA standards for "Class A" classification [16]. After composting, the biosolids were drum sieved before trucking to Farmington. The fly ash required no sieving.

Composted biosolids and fly ash treatments were applied to the nursery containers February 24, 2005 at two agricultural rates: 22.75 Mg ha^{-1} (82.1 g byproduct per container) and 45.5 Mg ha^{-1} (164 g byproduct per container). A third rate was applied based on two criteria: (1) the amount of Fe in the Farmington soil after a DTPA- (diethylenetriaminepentaacetic acid-) extractable Fe baseline soil test was conducted and (2) the percent available Fe in each amendment that could be applied to the soil as a fertilizer to correct a potential Fe deficiency. The complete baseline soil chemical analysis used to determine DTPA application rates was established earlier for both byproducts and is presented in Table 1. The DTPA Fe content of the Farmington soil was 1.2 mg kg^{-1}. For soils with a test report of 0.0–2.5 mg kg^{-1} Fe, Jones and Jacobsen [22] recommend an application rate of 4.5 kg Fe ha^{-1} in order to overcome Fe deficiency in susceptible crops. The DTPA-extractable iron in the fly ash was 0.00609% Fe and 0.0329% Fe for the biosolids. Expressed as fractions, DTPA values determined for each byproduct were used as divisors to the 4.5 kg Fe ha^{-1} recommendation which yielded the equivalent of 74 Mg of fly ash ha^{-1} (270.5 g applied per container) and 14 Mg biosolids ha^{-1} (50.1 g per container) application rates. These were the highest and lowest rates for fly ash and composted biosolids, respectively. Treatments were incorporated by removing the top 10 cm of soil from each nursery container, placing the contents into a plastic bucket, and mixing in the amendment before returning the contents to the nursery container. An Fe fertilizer check, Sprint Sequestrene 138 (use of a trademarked product does not imply an endorsement by the NMSU Agricultural Experiment Station), 6% EDDHA chelated Fe (Becker Underwood, Ames Iowa), was applied as a soil drench once at week three at an application rate of 4.5 kg Fe ha^{-1} (275 mg Fe EDDHA per container). The application rate was based on the soil test report and percent available Fe in the product (6%) to supply the literature recommendations as described previously. Unamended soil served as the control.

Table 1: Selected chemical properties of Albuquerque biosolids and APS fly ash

Characteristic	Composted biosolids	Fly ash[b]
pH (1 : 2)[c]	7.4	12.4
EC (dS m^{-1})[d]	14.0	6.7
SAR (mmol L^{-1})[d]	4.75	2.04
NO$_3$-N (mg kg^{-1})[e]	71.3	2.71
TKN (mg kg^{-1})[f]	1850.0	NT
P (mg kg^{-1})[c]	231.3	17.0
K (mg kg^{-1})[c]	5723.3	11.7
Zn (mg kg^{-1})[c]	44.9	0.6
Fe (mg kg^{-1})[c]	420.3	78.4
Fe by DTPA (mg kg^{-1})[z]	329.0	60.90

Mn (mg kg^{-1})[c]	20.6	8.1
Cu (mg kg^{-1})[c]	15.0	1.2
Ca (mg kg^{-1})[c]	3557.0	5650.0
Mg (mg kg^{-1})[c]	657.7	31.0
Na (mg kg^{-1})[c]	855.3	53.6
S (mg kg^{-1})[e]	529.5	306.7
Al (mg kg^{-1} [e]	807.2	348.7
As (mg kg^{-1})[e]	18.5	10.3
B (mg kg^{-1})[e]	40.1	59.5
Ba (mg kg^{-1})[e]	211.8	904.1
Be (mg kg^{-1})[e]	ND	ND
Cd (mg kg^{-1})[e]	2.3	1.3
Co (mg kg^{-1})[e]	3.4	0.8
Cr (mg kg^{-1})[e]	13.9	2.9
Mo (mg kg^{-1})[e]	ND	ND
Ni (mg kg^{-1})[e]	8.6	1.9
Pb (mg kg^{-1})[e]	18.8	5.0
Se (mg kg^{-1})[e]	ND	ND
Tl (mg kg^{-1})[e]	ND	ND
V (mg kg^{-1})[e]	22.1	8.4
Bi (mg kg^{-1})[e]	ND	ND
Li (mg kg^{-1})[e]	8.6	3.8
Sr (mg kg^{-1})[e]	163.8	37.9
Si (mg kg^{-1})[e]	322.7	308.7
Ag (mg kg^{-1})[e]	5.2	ND

[a]Mean of 6 samples for pH, P, K, Zn, Fe, Mn, Cu, Ca, Mg, and Na.

[b]Mean of 3 samples for S, Al, As, B, Ba, Be, Cd, Co, Cr, Mo, Ni, Pb, Se, Tl, V, Bi, Li, Sr, Si, and Ag.

[c]Analyzed at Soil Chemistry Research Laboratory, NMSU, Las Cruces, NM.

[d]Analyzed at Agricultural Testing and Research Laboratory, NAPI, Farmington, NM.

[e]Analyzed at the NMSU Soil, Water, and Air Testing Laboratory, Las Cruces, NM.

[Literature value supplied by Glass (personal communication, 2006).

ND = not detected. NT = not tested.

Plant Material

Hybrid poplar OP-367 (Populus deltoides × P. nigra) is a commercial hybrid that performs well in Farmington but benefits from supplemental Fe. Uniform 30 cm long cuttings obtained from Broadacres Nursery (Hubbard, OR) were soaked for 3 days in tap water before transplanting (February 27, 2005) directly into nursery containers.

Other Cultural Practices

Greenhouse temperatures averaged 16°C (min) and 41°C (max). Containers were kept at or below field capacity and were not leached to examine the potential for salt buildup in the soil. The total amount of water applied to each container over the course of the study was 647 mm.

The control soil has a low N content (less than 1% organic matter). In addition, N is volatilized during combustion, making fly ash even lower in N content. Therefore, the fly ash- and Fe EDDHA-treated trees received the equivalent of 90 kg N ha^{-1} N (split into 14 application times spread over the course of the study, applied to trees in the irrigation water) in the form of urea ammonium nitrate (UAN 32-0-0) in order to maintain similar values of N in all treatments. This was necessary because the composted biosolids contained 85.5 kg N ha^{-1} when applied at the 44.5 Mg ha^{-1} rate determined from NO_3-N using the ion-specific electrode method [23] and Kjeldahl N measured by the City of Albuquerque (Glass, 2006 personal communication) (Table 1).

Chlorophyll Analysis

Leaf chlorophyll content was monitored using a handheld Minolta SPAD- (soil plant analysis development-) 502 meter. The SPAD meter nondestructively measures transmittance of the leaf in red and infrared wavelengths (650 and 940 nm, resp.) giving a unitless leaf "greenness" value [24]. As SPAD values increase, leaf chlorosis decreases. For the clone OP-367, SPAD values were previously shown to correlate well with Fe (r^2=0.58) and total chlorophyll analyzed by HPLC (r^2=0.58) [25]. SPAD readings were made on April 12, and June 22 by measuring the first 10 fully expanded leaves (beginning 5-6 nodes down from the apical bud) on each tree.

Postharvest Analysis

The study was terminated July 6, 2005 at which point leaves were removed from each tree and passed through a leaf area meter. Leaves were decontaminated of Fe sources from dust/soil by dipping in a 0.01% phosphate-free detergent bath (0.1 mL detergent L^{-1} tap H_2O) [26] followed by rinsing with tap water under low pressure to remove soap residues. Leaves were then dipped into two baths of distilled water, bagged, dried for 24 hours at 70°C, and then weighed.

Stems were severed 2 cm from the top of the original cutting and measured for basal diameter and overall length. Soil was removed from the root ball (roots plus original cutting) then sieved through a 3 mm × 3 mm mesh to remove root pieces. Roots were then dipped in six water baths to remove residual soil. Roots were then severed from the original cutting and rinsed under low pressure. Stems and roots were then dried separately at 70°C for 72 hours before weighing.

Plant Fe and N Analysis

Dried leaves and stem material were ground to a fine powder using a stainless steel coffee grinder (cleaned thoroughly between samples) and stored in snap cap vials at room temperature until chemical analysis. Plant Fe was extracted with 20% trace metal grade HCl after dry ashing [27] and analyzed by inductively coupled plasma-optical emission spectroscopy (ICP-OES; Perkin-Elmer Optima 4300 DV ICP-OES). Plant total nitrogen (TN) was determined directly by combustion (LECO TruSpec CNS).

Plant Tissue Heavy Metal Analysis

Following the method described by Miller [28], microwave-assisted acid digestion using Teflon pressure digestion vessels was used to extract Cr, Pb, Se, As, Ag, Ba, and Cd from leaves. Acid digests were then analyzed by ICP-OES. All plant tissue macro elements, Fe, B, and heavy metals are expressed on a dry weight basis.

Soil Analysis

Soil was analyzed for pH (1 : 2, soil : water), extractable P, and Fe by ammonium bicarbonate-DTPA (1.0 mol L^{-1} NH$_4$HCO$_3$ + 0.005 mol L^{-1} DTPA at pH 7.6) [29]. Extracts were analyzed by ICP-OES. Soil NO$_3$-N was analyzed using the ion specific electrode method [23]. Electrical conductivity (EC) and sodium adsorption ratio (SAR) were measured on saturated paste extracts.Soil Cr, Pb, Se, As, Ag, Ba, and Cd concentrations were determined by ICP-OES following the USEPA 3051A [30] microwave-assisted acid digestion method for soil samples.

All laboratory analyses were conducted at the Navajo Agricultural Products Industry (NAPI) Agricultural Testing and Research Laboratory (Farmington, NM), NMSU Plant and Environmental Sciences Soils Research group laboratory (Las Cruces, NM), and the NMSU Soil, Water, and Air Testing (SWAT) laboratory.

Experimental Design and Statistical Analysis

The study was a randomized complete block design on two benches to compensate for temperature gradients within the greenhouse. There were eight containers per each treatment. Containers were redistributed once per week within blocks on the benches to help ensure that all trees received equal amounts of light exposure.

Analysis of variance was done in SAS (Cary, NC) using the PROC Mixed statement. All pairwise comparisons were made when significant differences were observed using Fisher's protected LSD at an alpha 0.05 level calculated by the method described by Littell et al. [31]. Concerning plant elements (Fe, N, B, and Ba), when significant accumulations of these elements were found in stems and leaves, the statistics were performed on the total plant accumulation (leaves + stems). In the case when no significant differences were detected in stems but were in leaves, only leaves are reported to simplify the data reporting. Correlation analysis using the PROC CORR command was performed to determine linear relationships between plant growth and environmental/plant toxicity parameters (i.e., pH, EC, SAR, and boron).

Codes for biosolid treatments are referenced the following way: ACB 22.75, ACB 44.5 (for Albuquerque composted biosolids at 22.75 Mg ha^{-1} and 44.5 Mg ha^{-1} application rates, resp.), ACB DTPA (biosolids applied at a rate based upon its DTPA extractable, plant-available Fe). Codes for fly ash plus urea ammonium nitrate fertilizer treatments are referenced as FA 22.75 + UAN, FA 44.5 + UAN, and FA DTPA + UAN (fly ash applied at a rate based upon its DTPA extractable, plant-available Fe). The Fe EDDHA plus urea ammonium nitrate fertilizer treatment is referenced as Fe EDDHA + UAN.

RESULTS AND DISCUSSION

Soil/Plant Nutrient Status and Tree Growth after Amending with Byproducts

Soil Fe, Chlorosis, and Tissue Fe Responses

Even though soil Fe in OP-367 trees receiving Fe EDDHA + UAN was no different than the control (Table 2), these trees had the highest mean SPAD values (41.2 SPAD units; p<0.0001) and highest plant Fe (38.3 mg kg^{-1}; p<0.0001; Figure 1). Fly ash + UAN increased soil Fe concentration 18–46%, though these increases were also no different from the control soil (Table 2). However, chlorophyll and plant Fe in fly ash + UAN treatments increased in leaves according to the application rate in the following manner: SPAD values were highest in the FA DTPA + UAN rate (36.9 SPAD units with a 64% increase in plant Fe), followed by the FA 44.5 Mg ha^{-1} + UAN rate (SPAD value 34.7 and 64% increase in plant Fe) and FA 22.75 Mg ha^{-1} + UAN rate (SPAD value 32.8 and 43% increase in plant Fe; Figure 1).

Table 2: Soil NO$_3$-N, P, and Fe a Doak sandy loam after amending with Fe EDDHA + UAN (urea ammonium nitrate), fly ash + UAN, and composted biosolids (n=8 per treatment). Means with the same letter are not significantly different at =0.05 level

Treatments[a]	Fe (mg kg^{-1})	NO$_3$-N (mg kg^{-1})	P (mg kg^{-1})
Control	4.44 d	2.3 c	9.39 d
Fe EDDHA + UAN	4.84 d	3.4 a	8.55 d
ACB DTPA	16.63 c	2.4 c	23.23 c
ACB 22.75	22.13 b	2.5 bc	30.71 b
ACB 44.5	37.09 a	3.3 ab	52.05 a
FA 22.75 + UAN	5.27 d	3.8 a	8.87 d
FA 44.5 + UAN	5.51 d	3.5 a	8.90 d
FA DTPA + UAN	6.48 d	3.7 a	8.59 d
Mean	12.80	3.1	18.79
LSD	3.28	0.82	2.56
F Value	104.09	4.64	306.1
Pr>F	<.0001	0.0004	<.0001

[a]Codes for biosolid treatments are referenced the following way: ACB 22.75, ACB 44.5 (for Albuquerque composted biosolids at 22.75 Mg ha^{-1} and 44.5 Mg ha^{-1} application rates, resp.), ACB DTPA (biosolids applied at a rate based upon its DTPA extractable, plant-available Fe; equivalent to 14 Mg ha^{-1}). Codes for fly ash plus urea ammonium nitrate fertilizer treatments are referenced as FA 22.75 + UAN, FA 44.5 + UAN, and FA DTPA + UAN (fly ash applied at a rate based upon its DTPA extractable, plant-available Fe; equivalent to 74 Mg ha^{-1}). The Fe EDDHA plus urea ammonium nitrate fertilizer treatment is referenced as Fe EDDHA + UAN.

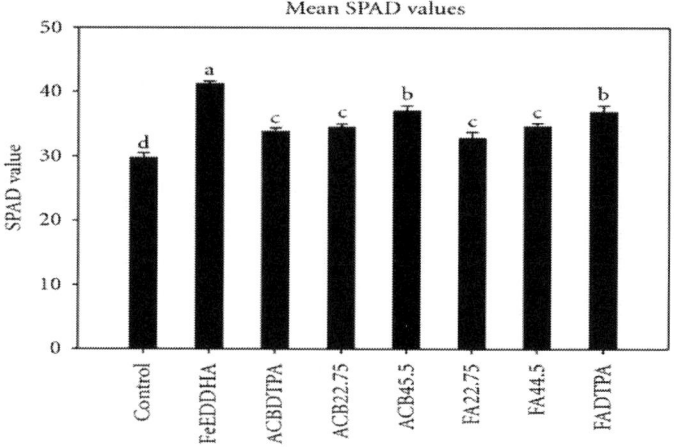

(a)

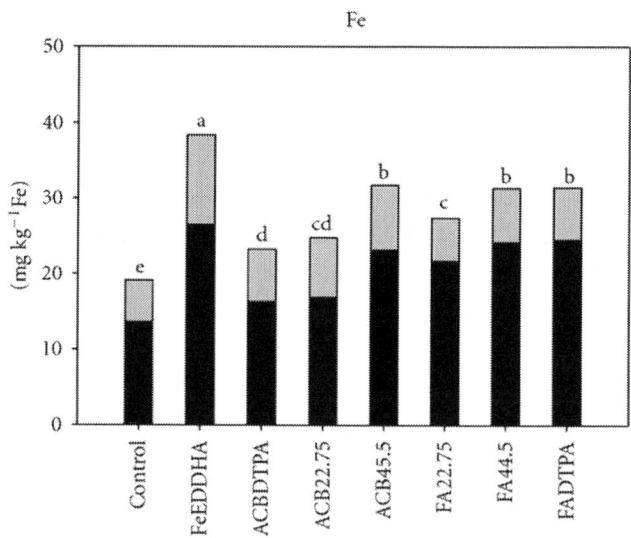

(b)

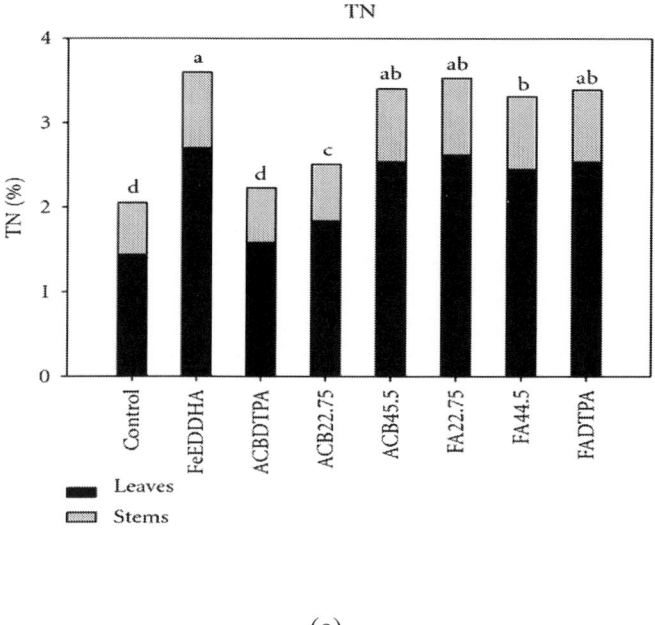

(c)

Figure 1: SPAD values (a), plant Fe (b), and total plant N (c) of the hybrid poplar clone OP-367 cultivated in soil amended with Fe ED-DHA + UAN (urea ammonium nitrate), fly ash + UAN, and composted biosolids. SPAD means are from two measurement periods. Means with the same letter are not significantly different at =0.05 level for combined leaf and stem material. Refer to Experimental Design and Statistical Analysis section for treatment codes.

The increases in plant Fe with the presumed boost in SPAD values were consistent with carrot when grown at a fly ash landfill site [32]. We have shown in previous studies that both Fe and N play a role in influencing SPAD values for the hybrid OP-367 [25]. Fly ash- and Fe EDDHA + UAN-treated poplars were given the same amount of UAN during the study, holding N constant. Acid-forming fertilizers containing NH_4^+, such as UAN, are known to lower the pH of the rhizosphere, making Fe and other microelements more plant-available in alkaline soils [33–35]. So the addition of UAN probably had a role in increasing plant Fe and SPAD values in the Fe EDDHA- and fly ash-treated trees.

Biosolids, on the other hand, significantly improved soil Fe contents by 275–700% above the control soil (P<.0001; Table 2), a response consistent with other studies when wastewater has been treated with FeCl salts and the subsequent biosolids are applied to calcareous soils [6, 11]. These trees also had improved leaf greenness and plant Fe (leaves plus stems) in the following way: the ACB DTPA (14 Mg ha^{-1}) gave SPAD values of 33.8 and a 22% plant Fe improvement while the ACB 22.75 Mg ha^{-1} treatment raised SPAD values to 34.5 and plant Fe by 30% above control trees (Figure 1). Composted biosolids applied at the 44.5 Mg ha^{-1} rate resulted in a 66% increase in plant Fe above the control (P<.0001). Although statistically lower than the Fe EDDHA+UAN treated trees, SPAD values in the ACB 44.5 Mg ha^{-1} trees increased to 37.0, 25% increase in leaf greenness above control trees (P<.0001; Figure 1).

Nitrogen and phosphorus bear some mention because, from a producer standpoint, both elements can be expensive farm inputs. Both elements are inherently low in fly ash [36] but can be quite elevated in biosolids. Because UAN applications were held constant for all fly ash and Fe EDDHA trees, as expected, all trees that received UAN had equal and higher soil NO_3-N when compared to control soil (P<.0004; Table 2). Except for the FA 44.5 rate, total leaf N was also equal and highest for trees receiving UAN (Figure 1; P<.0001). Only the ACB 44.5 rate equaled this response, increasing NO_3-N by 43% (Table 2) and plant N by 70% (Figure 1). The two lower composted biosolid rates were similar to the control for soil NO_3-N, but total plant N significantly increased by 27% in ACB-22.75-treated trees (Figure 1). Phosphorus contents of fly ash- and Fe EDDHA-treated soil were no different than the control (Table 2) because these trees received no supplemental P. On the other hand, P was 147–450% higher in soil treated with composted biosolids because Albuquerque uses iron chloride salts to remove P from the waste stream during the treatment process. This fact had some effect on soil salinity (more below).

Tree Growth

The following tree growth measurements were unaffected by the treatments: stem diameters (mean 10 mm), root dry weights (mean 9.4 mg kg^{-1}), total above-ground dry weights (mean 39.9 mg kg^{-1}), and root-to-above ground-biomass ratios (mean 0.31; data not shown).

Table 3 presents leaf area, leaf weight, stem weight, root weight, stem length, and stem diameter growth results. All treatments had greater leaf areas (P=.0020) except for the FA 22.75 Mg ha⁻¹, which compared equally to control trees (Table 3). The greatest leaf areas were from trees treated at the highest two fly ash application rates, followed by trees treated at the ACB 22.75 Mg ha⁻¹rate. For leaf dry weight, composted biosolids at the DTPA (14 Mg ha⁻¹) and 22.75 Mg ha⁻¹ rates had the greatest response followed by the FA 44.5 and FA DTPA + UAN treatments (P<.0397; Table 3); the Fe EDDHA + UAN and ACB 44.5 Mg ha⁻¹ rates were no different than the control. For stem dry weight, biosolids at all rates had the greatest response compared to the control trees (P= 0133); the control, Fe EDDHA, and fly ash plus UAN-treated trees were no different from one another (Table 3). Stem lengths were similar among the control soil, Fe EDDHA + UAN, composted biosolids, and FA 44.5 Mg ha⁻¹ trees while the FA 22.75 + UAN and FA DTPA (74 Mg ha⁻¹) rate had a nearly 2 and 4% reduction when compared to control tress (P=.0263; Table 3).

Table 3: Biomass results of OP-367 (P. deltoides × P. nigra) showing significant differences after amending with industrial byproducts. Means with the same letter are not significantly different at the α=0.05 level

	Leaf area	Leaf dry weight	Stem dry weight	Stem length
Treatments[a]	(cm²)	(g)	(g)	(cm)
Control	1419 d	15.0 b	14.0 d	124.6 abc
Fe EDDHA + UAN	1552 bc	15.1 b	14.5 bcd	124.0 abc
ACB DTPA	1563 bc	16.4 a	15.6 abc	127.6 ab
ACB 22.75	1584 abc	16.4 a	16.1 a	129.4 a
ACB 44.5	1464 cd	15.2 b	15.6 abc	126.3 ab
FA 22.75 + UAN	1527 bcd	15.2 b	14.3 cd	122.5 bc
FA 44.5 + UAN	1694 a	16.1 ab	14.8 bcd	124.2 abc
FA DTPA + UAN	1590 ab	15.3 ab	14.3 cd	119.5 c
LSD	121	1.1	1.3	5.5
F Value	3.79	2.3	2.84	2.5
Pr >F	0.002	0.0397	0.0133	0.0263

[a]Refer to Experimental Design and Statistical Analysis section for treatment codes.

Many complex factors influenced growth. Stem weights and stem lengths benefited from the added soil Fe while stem weights, stem lengths, and, to a lesser extent, stem diameters did not seem to benefit from the added NO_3-N (Table 4). Increasing the soil salinity, pH, and leaf B content also contributed to reduced growth (more below).

Table 4: Correlation matrix for 2005 Greenhouse Study showing growth versus soil and foliar parameter.Note: correlation coefficients (r values) are followed by P values; **indicates significance P<.05, ***indicates significance P<.001

	Leaf area	Leaf Wt.	Stem Wt.	Root Wt.	Stem Lnth.	Stem Dia.
Fe soil	−0.08	0.09	0.42***	−0.11	0.33**	0.11
	0.5197	0.4885	0.0006	0.3987	0.0068	0.4026
NO3-N	−0.15	−0.21	−0.45***	−0.08	−0.46***	−0.28**
	0.2364	0.0983	0.0002	0.5246	0.0001	0.0234
pH	−0.04	−0.16	−0.54***	−0.18	−0.41***	−0.14
	0.7443	0.1990	<0.0001	0.1481	0.0007	0.2756
SAR	0.16	−0.04	−0.35**	−0.14	−0.26**	−0.16
	0.2213	0.7499	0.0047	0.2592	0.0389	0.1945
EC	−0.14	0.08	0.24	−0.23	0.27**	−0.04
	0.2900	0.5548	0.052	0.0656	0.0279	0.7498
B leaves	0.21	−0.08	−0.20	0.14	−0.44***	0.03
	0.0982	0.5331	0.1198	0.2605	0.0003	0.7922

Environmental Considerations

Soil Sodium Adsorption Ratio, Electrical Conductivity, and pH

Salinity is a concern in our region because we average approximately 200 mm of rainfall per year, which equates to a low salt leaching potential. Any amendment containing high amounts of soluble salts poses the risk of increasing sodic/saline soil conditions. The sodium adsorption ratio (SAR) measures the proportion of Na^+ ions compared to the concentration of calcium Ca^{2+} plus Mg^{2+} in the saturated paste extract (the higher the SAR value, the more that Na^+ is dominating the soil chemistry). Electrical conductivity (EC) measures total soluble salt content (which can include NaCl but also N, P, Ca, Mg, and other fertilizer salts). An EC above 4 is generally considered the threshold point at which most agricultural crops suffer reduced yields [37].

When compared to the control, SAR values were similar for all treatments except for the ACB 44.5 (4.95 mmol L^{-1}), which experienced a 12% reduction in values from the control soil (5.65 mmol L^{-1}; P<0.0001; Table 5). This is explained by the additional Mg and Ca contributions to the soil from the parent biosolids material (Table 1). All of the SAR values were considerably below 13–15 mmol L^{-1} which is considered sodic and problematic for agricultural soils. Still, an inverse association was found between increasing SAR values and stem weight (r=-0.35; P=.0047) and stem length (r=-.026; P=.0389; Table 4). Within plants, the ionic balance of Ca, Mg, Na, components of the sodium adsorption ratio in soil, are known to be influenced by nitrogen fertilizer source [38]. The UAN may have had an effect on influencing the components of the SAR test and tree growth given the inverse relationship between NO_3-N versus growth parameters shown in Table4.

Table 5: Soil sodium adsorption ratio (SAR), electrical conductivity (salinity), and pH of a Doak sandy loam amended with industrial byproducts.

Treatments[a]	SAR (mmol L^{-1})	EC (dS m^{-1})	pH saturated paste
Control	5.65 ab	3.24 d	8.62 b
Fe EDDHA + UAN	5.78 a	3.17 d	8.66 ab
ACB DTPA	5.42 b	3.47 c	8.58 bc
ACB 22.75	5.55 ab	3.77 b	8.50 cd
ACB 44.5	4.95 c	4.04 a	8.44 d
FA 22.75 + UAN	5.73 a	3.10 de	8.64 b
FA 44.5 + UAN	5.62 ab	3.16 d	8.67 ab
FA DTPA + UAN	5.67 a	3.03 e	8.76 a
Mean	5.54	3.37	8.6
LSD	0.24	0.18	0.1
F Value	9.32	31.71	7.33
Pr >F	<0.0001	<0.0001	<0.0001

[a]Refer to Experimental Design and Statistical Analysis section for treatment codes.

The Fe EDDHA, FA 22.75, and FA 44.5 + UAN treatments were similar to control soil for EC. However, soil treated at the FA DTPA + UAN (74 Mg ha^{-1}) had a significant decrease in conductivity compared to control soil for reasons unknown. On the other hand, composted biosolid-treated soil increased EC by 21–41% above the control soil (P<.0001; Table 5). The increase in EC was expected because biosolids are high in soluble salts. In fact, at the ACB 44.5 Mg ha^{-1} rate, EC reached 4 dS m^{-1}. These levels were below the 5.5 dS m^{-1}tolerance limit defined for hybrid poplar [39] and did not appear to affect above-ground growth negatively (i.e.r=0.27, P=.0279; for the relationship between EC and stem length; Table 4). Although no significant relationship was demonstrated between EC and root weight (r=-0.23; P=.0656; Table 4), the potential for decreasing this parameter with increasing salinity exists given that containers were not leached. In later field plot studies, we found no salinity increases in plots amended with composted biosolids at 44 Mg ha^{-1} [40]. In the latter study, a total of 983 mm of water (irrigation + rainfall) was applied during the second growing season alone, which provided sufficient leaching potential; all biosolids field plots never exceeded an EC of 1 dS m^{-1}when sampled at a depth of 30 cm.

Another concern is applying an amendment that may have the potential to raise soil pH in already calcareous conditions. The pH ranged from 8.4 in soil treated at the ACB 22.75 Mg ha^{-1} rate to 8.8 in the FA DTPA + UAN-treated soil (equivalent of 74 Mg ha^{-1}); the control soil had a pH of 8.6 (P<.0001; Table 5). Low S-containing western US lignite coals typically produce alkaline ash [9, 41], which explains the pH increase in accordance with increasing application rate of fly ash. The pH increase was related to a reduction in stem weights (r=-0.54; P<.0001) and stem length (r=-0.47; P=.0007), and the general trend was that as pH increased, leaf area, leaf weight, root weight, and stem diameter decreased (Table 4).

Composted biosolids, when applied at the 22.75 Mg ha^{-1} rate, decreased pH below the control soil to 8.4. As salt concentration increases, soluble cations, such as Ca^{2+} and Mg^{2+}, replace acidic exchangeable cations ($H^+$and Al^{3+}) in the soil solution, lowering the pH of the soil extract solution [35, 43, 44]. The reduction in pH was probably related to the soluble salt content of the material. In field studies with the same biosolid application rates as in this study, under a leaching fraction, soil pH was equal among biosolid treatment [40].

Other Environmental Considerations

The following metals were analyzed in soils from acid digests: Cr, As, Ag, Se, Pb, Cd, and Ba. Arsenic, Ag, and Se were not detected in either soils or plants. Lead, Cr, and Cd also did not increase in soils in either study, averaging 7.71 mg kg^{-1} for Cr, 5.08 mg kg^{-1} for Pb, and 0.16 mg kg^{-1} for Cd (Table 6). These levels were below USEPA (Part 503 Rule) and European Union (Directive 86/278/EEC) regulations for heavy metal loading rates for biosolids applied to agricultural lands [42].

Table 6: Soil Cr, Pb, Ba, and Cd levels in a Doak sandy loam amended with industrial byproducts

	Cr	Pb	Ba	Cd
	(mg kg^{-1})	(mg kg^{-1})	(mg kg^{-1})	(mg kg^{-1})
Literature				
U.S. EPA 40 CFR 503 Rule[a]	—	300.0	—	39.00
European Union limit values[b]	—	750–1,200	—	20–40
EU proposed	1000.0	750.0	—	10.00
Treatments				
Control	7.28 a	5.19 a	97.23 e	0.16 a
Fe EDDHA + UAN	7.60 a	5.19 a	98.18 de	0.17 a
ACB DTPA	8.34 a	5.07 a	105.83 cd	0.16 a
ACB 22.75	7.80 a	5.05 a	102.33 de	0.15 a
ACB 44.5	7.92 a	5.12 a	100.84 de	0.17 a
FA 22.75 + UAN	7.68 a	5.22 a	111.48 bc	0.17 a
FA 44.50 + UAN	7.51 a	4.93 a	119.25 b	0.17 a
FA DTPA + UAN	7.53 a	4.86 a	137.63 a	0.17 a
Mean	7.71	5.08	109.10	0.16
LSD	NS	NS	8.46	NS
F Value	0.34	1.02	20.93	0.95
Pr >F	0.9302	0.4274	<.0001	0.48

[a]Pollutant concentration limits and loading rates for land application in the United States.

[b]European Union limit values for concentrations of heavy metals in biosolids for use on land [42]

Refer to Experimental Design and Statistical Analysis section for treatment codes.

Boron and Ba, however, did present potential environmental concerns. Boron increased in the leaves of fly ash + UAN amended poplars by 23% (22.75 Mg ha^{-1} rate), 45% (44.5 Mg ha^{-1} rate) to 85% (rate equivalent to 74 Mg ha^{-1}) (Figure 2). Although B is a micronutrient needed by plants in trace amounts, toxicity symptoms and decreased crop yields result from the application of unweathered fly ash [9, 41, 45, 46]. At the highest fly ash application rate, leaf B reached 93.6 mg kg^{-1}, which began to approach toxicity levels (above 141 mg B kg^{-1}dwt) defined for OP-367 by Bañuelos et al. [47]. Indeed, the elevated B levels found in the leaves of fly ash-treated trees inversely correlated with stem lengths (r=-0.44; P=.0003; Table 4). As B moves easily with irrigation waters, accumulation of B may have been mitigated if the containers were leached. Thus, boron accumulations to toxic levels present an environmental concern for agricultural land application of fly ash to our soils if not leached regularly and managed carefully.

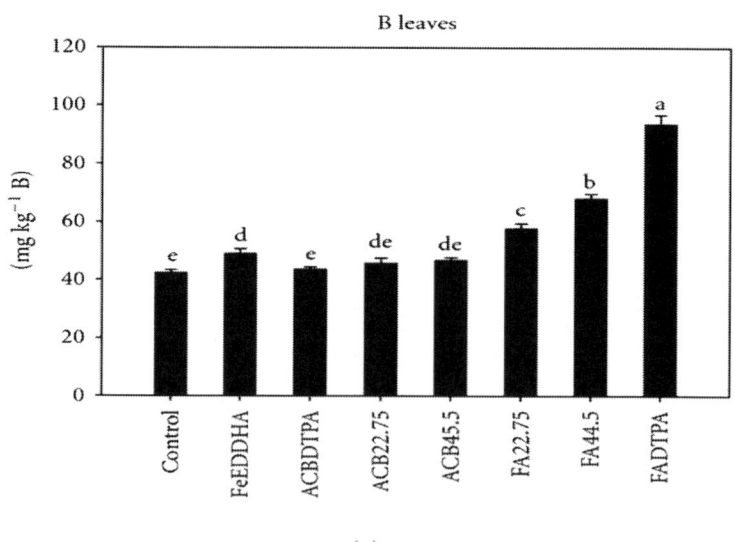

(a)

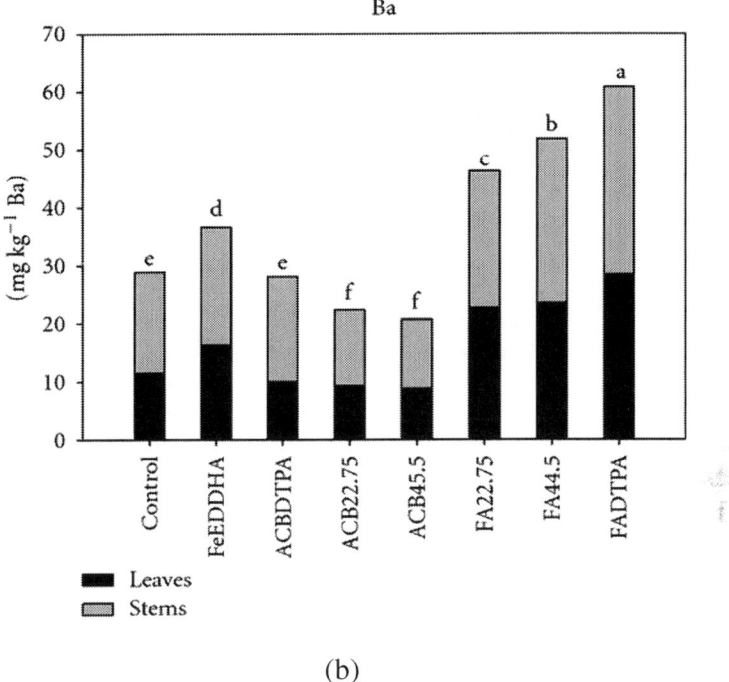

(b)

Figure 2: Leaf B (a) and leaf + stem Ba (b) of the hybrid poplar clone OP-367 cultivated in soil amended with Fe EDDHA + UAN (urea ammonium nitrate), fly ash + UAN, and composted biosolids. Means with the same letter are not significantly different at α=0.05 level for combined leaf and stem material. Refer to Experimental Design and Statistical Analysis section for treatment codes.

Barium increased 7–42% in fly ash-amended soil (Table 6). As a consequence, the stem and leaf Ba levels combined increased 60–110% (46.3–60.8 mg kg^{-1}) compared to control trees (28.9 mg kg^{-1}; P<.0001; Figure2). The increase of Ba is not uncommon when fly ash is applied to land [46]. With regards to Ba, it is difficult to assume environmental safety because we did not analyze for potentially toxic forms (barium carbonate, barium chloride, and barium acetate) [48].

As for the composted biosolids material or the composted biosolid/soil mixtures, we did not analyze for the presence of antibiotics and/or personal care products. Recent attention to these persistent organic constituents shows that biosolids, especially when not composted, may impart potential antibiotic resistance in soil micro-organisms or molecules from personal care products may exhibit hormonal effects on aquatic organisms [14,49]. Although the risk factors associated with antibiotics and personal care products can be mitigated by composting [49, 50], an analysis for these constituents is essential to strengthening our knowledge of the material used in this study and in developing a future comprehensive environmental risk assessment when composted biosolids might be considered for agricultural land application.

CONCLUSIONS

The Fe EDDHA + UAN had the highest SPAD values and plant Fe (stem + leaves). This is an expensive synthetic fertilizer input with 6% plant-available iron. In comparison, fly ash plus UAN significantly increased SPAD values and plant Fe in hybrid poplars even though soil Fe remained statistically similar to the control soil. The trend in SPAD values and plant Fe generally followed fly ash application rate—the higher the rate, the greater the response. Despite the increase in soil pH from fly ash additions, which would imply even lower solubility of micronutrients, and the fact that N from UAN remained constant among the fly ash (and Fe EDDHA treatments), the uptake of Fe in fly ash-amended poplars was likely related to an acidulation of the rhizosphere from UAN applications which made Fe more available just at the root/soil interface. Leaf area was greatest in hybrid poplar trees grown in soil amended with fly ash treatments, but growth was similar to the control for all other biometric parameters. The highest plant B accumulations occurred in the fly ash + UAN treatments, especially at the fly ash rate equivalent to $74\,Mg\,ha^{-1}$. Increasing leaf B was inversely associated with stem length in these trees. Stem and leaf Ba was also highest in trees grown in fly ash-treated soil. Potentially toxic forms of Ba found in fly ash were not measured, which warrants further investigation. Finally, the fact that NO_3-N was inversely related to growth raises the need for us to conduct a UAN exclusion study before a definitive recommendation can be made concerning fly ash applications to agricultural lands.

Biosolids significantly increased soil Fe and P in all treatments, and NO_3-N only at the $44.5\,Mg\,ha^{-1}$ rate. Although lower than the Fe EDDHA treatment, composted biosolid-treated trees had SPAD and plant Fe values significantly increase in proportion to application rate, showing that composted biosolids could supply plant-available Fe to trees growing on an alkaline soil. Likewise, total plant N increased, but only the 44.5 Mg ha^{-1} treatment equaled the response of UAN-treated trees. Trees grown under composted biosolids applied at the DTPA (14 Mg ha^{-1}) and 22.75 Mg ha^{-1} rates generally had the highest growth (stem dry weight, stem length, and leaf dry weight). Saturated paste extracts of the byproducts demonstrated that composted biosolids had the most potential for increasing soil salinity due to their complex mixtures of soluble salts. Soil treated with the 44.5 Mg ha^{-1} rate had an EC of 4 dS m^{-1} which may explain why the lower biosolids application rates generally had greater tree growth. An inverse relationship between increasing salinity and decreasing root dry weight was shown, but salinity increases were below tolerance levels defined for hybrid poplar, and salts would be flushed from the root zone if leaching of containers was allowed in this study.

Overall, trees grown under biosolids generally exhibited the greatest response in regards to growth, soil, and plant Fe, N, and P increases without the need to provide supplemental N nutrition in the form of UAN. Other benefits not reported included an increase in Zn, Cu, and Mn in trees grown with composted biosolid amended soil. Amendment/soil mixtures showed little potential for environmental hazard in terms of heavy metal increases. It appears that a onetime application of the ACB 22.75 Mg ha^{-1} rate is sufficient to supply plant-available Fe and growth benefits to hybrid poplar seedlings without the risk of increasing salinity inunleached circumstances. The beneficial recycling of nutrients from biosolids to agricultural crops produced on a calcareous soil is feasible. An analysis of the Albuquerque composted biosolids for antibiotics and personal care products would add value to future field plot studies.

ACKNOWLEDGMENTS

The authors would like to thank the U.S. Department of Energy National Energy Technology Laboratory and the Combustion Byproducts Recycling Consortium for partial funding of this paper. The authors wish to thank Steven Glass, City of Albuquerque, for donating the composted biosolids and Bruce Salisbury, Arizona Public Services, for the fly ash.

REFERENCES

1.   R. H. Loeppert, L.-C. Wei, and W. R. Ocumpaugh, "Soil factors influencing the mobilization of iron in calcareous soils," in Biochemistry of Metal Micronutrients in the Rhizosphere, J. A. Manthey, D. E. Crowley, and D. G. Luster, Eds., pp. 343–360, CRC Press, Boca Raton, Fla, USA, 1994.

2.   B. L. Black and R. H. Zimmerman, "Mixtures of coal ash and compost as substrates for highbush blueberry," Journal of the American Society for Horticultural Science, vol. 127, no. 5, pp. 869–877, 2002.

3.   M. A. Cavaleri, D. W. Gilmore, M. Mozaffari, C. J. Rosen, and T. R. Halbach, "Hybrid poplar and forest soil response to municipal and industrial by products: a greenhouse study," Journal of Environmental Quality, vol. 33, no. 3, pp. 1055–1061, 2004.

4.   R. F. Korcak, "Utilization of coal combustion by-products in agriculture and horticulture," inAgricultural Utilization of Urban and Industrial By-Products, pp. 107–130, American Society of Agronomy: Crop Science Society of America: Soil Science Society of America, Madison, Wis, USA, 1995.

5.   P. U. Sashee Kumar and S. Sambath Kumar, "Fly ash: potential applications," Chemical Engineering World, vol. 37, no. 2, pp. 83–85, 2002.

6.   R. Moral, J. Moreno-Caselles, M. Perez-Murcia, and A. Perez-Espinosa, "Improving the micronutrient availability in calcareous soils by sewage sludge amendment," Communications in Soil Science and Plant Analysis, vol. 33, no. 15–18, pp. 3015–3022, 2002.

7.   P. Parkpian, S. T. Leong, P. Laortanakul, and J. Juntaramitree, "An environmentally sound method for disposal of both ash and sludge wastes by mixing with soil: a case study of Bangkok plain,"Environmental Monitoring and Assessment, vol. 74, no. 1, pp. 27–43, 2002.

8.   American Coal Ash Association, "Glossary of terms concerning the management and use of coal combustion products (CCPs)," 2003, http://www.acaa-usa.org/.

9. C. L. Carlson and D. C. Adriano, "Environmental impacts of coal combustion residues," Journal of Environmental Quality, vol. 22, no. 2, pp. 227–247, 1993.

10. G. W. Dickerson, A Sustainable Approach to Recycling Urban and Agricultural Organic Wastes, New Mexico State University Cooperative Extension Service, Las Cruces, NM, USA, 2000.

11. G. M. Zinati, Y. Li, and H. H. Bryan, "Accumulation and fractionation of copper, iron, manganese, and zinc in calcareous soils amended with composts," Journal of Environmental Science and Health Part B, vol. 36, no. 2, pp. 229–243, 2001.

12. C. M. Cooke, L. Gove, F. A. Nicholson, H. F. Cook, and A. J. Beck, "Effect of drying and composting biosolids on the movement of nitrate and phosphate through repacked soil columns under steady-state hydrological conditions," Chemosphere, vol. 44, no. 4, pp. 797–804, 2001.

13. K. Bousselhaj, S. Fars, A. Laghmari, A. Nejmeddine, N. Ouazzani, and C. Ciavatta, "Nitrogen fertilizer value of sewage sludge co-composts," Agronomie, vol. 24, no. 8, pp. 487–492, 2004.

14. W. F. Jaynes and R. E. Zartman, "Origin of talc, iron phosphates, and other minerals in biosolids," Soil Science Society of America Journal, vol. 69, no. 4, pp. 1047–1056, 2005.

15. BNR Operation in Wastewater Treatment Plants Task Force, Biological Nutrient Removal Operation in Wastewater Treatment Plants, Water Environment Federation and American Society of Civil Engineers and Environmental and Water Resources Institute, Eds., McGraw-Hill, New York, NY, USA, 2005.

16. City of Albuquerque, "Compost Facility Operations," 2008,http://www.abcwua.org/content/view/197/350/.

17. A. Dellantonio, W. J. Fitz, H. Custovic et al., "Environmental risks of farmed and barren alkaline coal ash landfills in Tuzla, Bosnia and Herzegovina," Environmental Pollution, vol. 153, no. 3, pp. 677–686, 2008.

18. R. J. Haynes, "Reclamation and revegetation of fly ash disposal sites—challenges and research needs,"Journal of Environmental Management, vol. 90, no. 1, pp. 43–53, 2009.

19. WU. Chenxi, A. L. Spongberg, and J. D. Witter, "Determination of the persistence of pharmaceuticals in biosolids using liquid-

chromatography tandem mass spectrometry," Chemosphere, vol. 73, no. 4, pp. 511–518, 2008.

20.   J. A. Ippolito, K. A. Barbarick, M. W. Paschke, and R. B. Brobst, "Infrequent composted biosolids applications affect semi-arid grassland soils and vegetation," Journal of Environmental Management, vol. 91, no. 5, pp. 1123–1130, 2010.

21.   C. W. Keetch, Soil survey of San Juan County New Mexico: Eastern Part.: USDA SCS, USDA BIA and BOR, NMSU Agricultural Experiment Station, 1980.

22.   C. Jones and J. Jacobsen, "Micronutrients: cycling, testing and fertilizer recommendations," in Nutrient Management, a Self-Study Course from the MSU Extension Service Continuing Education Series, Montana State University, 2003.

23.   W. C. Dahnke, "Use of the nitrate specific ion electrode in soil testing," Communications in Soil Science and Plant Analysis, vol. 2, no. 2, pp. 73–84, 1971.

24.   J. S. Schepers, T. M. Blackmer, and D. D. Francis, "Chlorophyll meter method for estimating nitrogen content in plant tissue," in Handbook of Reference Methods for Plant Analysis, Y. P. Kalra, Ed., pp. 129–134, CRC Press, Boca Raton, Fla, USA, 1998.

25.   K. Lombard, M. O'Neill, J. Mexal et al., "Can soil plant analysis development values predict chlorophyll and total Fe in hybrid poplar?" Agroforestry Systems, vol. 78, no. 1, pp. 1–11, 2009.

26.   C. R. Campbell and C. O. Plank, "Preparation of plant tissue for laboratory analysis," in Handbook of Reference Methods for Plant Analysis, Y. P. Kalra, Ed., pp. 37–49, CRC Press, Boca Raton, Fla, USA, 1998.

27.   D. E. Baker, G. W. Gorsline, C. G. Smith, W. I. Thomas, W. E. Grube, and J. L. Ragland, "P, K, Ca, Mg, Na, B, Zn, Mn, Fe, Cu, and Mo of botanical materials (dry ash)," Agronomy Journal, vol. 56, pp. 133–136, 1964.

28.   R. O. Miller, "Microwave digestion of plant tissue in a closed vessel," in Handbook of Reference Methods for Plant Analysis, Y. P. Karla, Ed., pp. 69–73, CRC Press, New York, NY, USA, 1998.

29.   P. N. Soltanpour and A. P. Schwab, "A new soil test for simultaneous extraction of macro-and micro-nutrients in alkaline soils," Communications in Soil Science and Plant Analysis, vol. 8, pp. 195–207, 1977.

30. USEPA, Microwave assisted acid digestion of sediments, sludges, soils, and oils, US Environmental Protection Agency, Washington, DC, USA, 1998.

31. R. C. Littell, W. W. Stroup, and R. J. Freund, SAS for Linear Models, SAS Institute, Cary, NC, USA, 4th edition, 2002.

32. L. H. Weinstein, M. A. Arthur, R. E. Schneider, et al., "Uptake of chemical elements by terrestrial plants growing on a coal fly ash landfill," in Trace Elements in Coal and Coal Combustion Residues, R. F. Keefer and K. S. Sajwan, Eds., pp. 213–237, CRC Press, Boca Raton, Fla, USA, 1993.

33. V. D. Fageria, "Nutrient interactions in crop plants," Journal of Plant Nutrition, vol. 24, no. 8, pp. 1269–1290, 2001.

34. H. M. Reisenauer, "The interactions of manganese and iron," in Biochemistry of Metal Micronutrients in the Rhizosphere, J. A. Manthey, D. E. Crowley, and D. G. Luster, Eds., pp. 147–164, CRC Press, Boca Raton, Fla, USA, 1994.

35. S. L. Tisdale, W. L. Nelson, and J. D. Beaton, Soil Fertility and Fertilizers, Macmillan Publishing Company, New York, NY, USA, 4th edition, 1985.

36. D. K. Gupta, U. N. Rai, R. D. Tripathi, and M. Inouhe, "Impacts of fly-ash on soil and plant responses," Journal of Plant Research, vol. 115, no. 6, pp. 401–409, 2002.

37. E. V. Maas and G. J. Hoffman, "Crop salt tolerance-current assessment," Journal of the Irrigation and Drainage Division, vol. 103, pp. 115–134, 1977.

38. E. A. Kirkby and K. Mengel, "Ionic balance in different tissues of the tomato plant in relationship to nitrate, urea, or ammonium nutrition," Plant Physiology, vol. 42, pp. 6–14, 1967.

39. M. C. Shannon, G. S. Bañuelos, J. H. Draper, H. Ajwa, J. Jordahl, and L. Licht, "Tolerance of hybrid poplar (Populus) trees irrigated with varied levels of salt, selenium, and boron," International Journal of Phytoremediation, vol. 1, no. 3, pp. 273–288, 1999.

40. K. Lombard, M. O'Neill, R. Heyduck et al., "Composted biosolids as a source of iron for hybrid poplars (Populus sp.) grown in northwest New Mexico," Agroforestry Systems, vol. 81, no. 1, pp. 45–56, 2011.

41. M. P. Menon, K. S. Sajwan, G. S. Ghuman, J. James, and K. Chandra, "Elements in coal and coal ash residues and their potential for agricultural crops," in Trace Elements in Coal and Coal Combustion Residues, R. F. Keefer and K. S. Sajwan, Eds., pp. 259–287, CRC Press, Boca Raton, Fla, USA, 1993.

42. National Research Council, "Committee on Toxicants and Pathogens in Biosolids Applied to Land," inBiosolids Applied to Land: Advancing Standards and Practices, National Research Council, Board on Environmental Studies and Toxicology, Washington DC, USA, 2002.

43. A. S. Al-Busaidi and P. Cookson, "Salinity-pH relationships in calcareous soils," Agricultural and Marine Sciences, vol. 8, no. 1, pp. 41–46, 2003.

44. NRCS, "Use of reaction (pH) in soil taxonomy," 1993,http://soils.usda.gov/technical/technotes/note8.html

45. W. A. Dick, L. Chen, and Y. Hao, "Beneficial uses of clean coal combustion by-products: soil amendment and coal refuse treatment examples and case studies," 2000,http://www.mcrcc.osmre.gov/MCR/Resources/ccb/PDF/Use_and_Disposal_of_CCBs_at_Coal_Mines.pdf#page=155

46. A. P. Schwab, "Extractable and plant concentrations of metals in amended coal ash," in Trace Elements in Coal and Coal Combustion Residues, R. F. Keefer and K. S. Sajwan, Eds., pp. 185–211, CRC Press, Boca Raton, Fla, USA, 1993.

47. G. S. Bañuelos, M. C. Shannon, H. Ajwa, J. H. Draper, J. Jordahl, and L. Licht, "Phytoextraction and accumulation of boron and selenium by poplar (Populus) hybrid clones," International Journal of Phytoremediation, vol. 1, no. 1, pp. 81–96, 1999.

48. E. Apedaile, "CH2M Hill Canada Limited, and D. Cole, 2002. Health aspects of biosolids land application," A report prepared for the City of Ottawa: Ottawa, Canada.

49. K. Xia, A. Bhandari, K. Das, and G. Pillar, "Occurrence and fate of pharmaceuticals and personal care products (PPCPs) in biosolids," Journal of Environmental Quality, vol. 34, no. 1, pp. 91–104, 2005.

50. H. Dolliver, S. Gupta, and S. Noll, "Antibiotic degradation during manure composting," Journal of Environmental Quality, vol. 37, no. 3, pp. 1245–1253, 2008.

Chapter 5

Nanoenhanced Materials for Reclamation of Mine Lands and Other Degraded Soils: A Review

Ruiqiang Liu and Rattan Lal

Carbon Management & Sequestration Center, School of Environment and Natural Resources, Ohio State University, 210 Kottman Hall, 2021 Coffey Road, Columbus, OH 43210, USA

ABSTRACT

Successful mine soil reclamation facilitates ecosystem recovery, minimizes adverse environmental impacts, creates additional lands for agricultural or forestry uses, and enhances the carbon (C) sequestration. Nanoparticles with extremely high reactivity and deliverability can be applied as amendments to improve soil quality, mitigate soil contaminations, ensure safe land–application of the conventional amendment materials (e.g., manures and biosolids), and enhance soil erosion control. However, there is no report on using nanoenhanced materials for mine soil reclamation. Through reviewing the up-to-date research results on using environment-friendly nanoparticles for agricultural soil quality improvement and for contaminated soil remediation, this paper synthesizes that these nanomaterials with high potentials for mine soil reclamation include zeolites, zero-valent iron nanoparticles, iron oxide nanoparticles, phosphate-based nanoparticles, iron sulfide nanoparticles and C nanotubes. Transport of these particles in the environment and their possible ecotoxicological effects are also discussed. Additionally, this article proposes a practical and economical approach to applying nanotechnology for mine soil reclamation: adding small amounts of nanoparticles to the conventional soil amendment materials and then applying the mixtures for soil quality improvements. Hence the cost of using nanoparticles is reduced and the benefits of both nanoparticles and the conventional amendment materials are harnessed.

INTRODUCTION

Ever since the commencement of industrial-scale mining of coal and other minerals, drastic environmental impacts have been recorded arising from both the mined lands and from the wastes left behind at the surface [1, 2]. The local landscape and the soil quality are among the most severely disturbed environmental components by the mining processes through directly stripping the vegetation and soil layers (open-pit mining) and/or through depositing the ores and mining wastes on the soils [3]. Dramatic alterations of the geological environment of the coal/ores and the mining wastes significantly reduce the chemical stability of the minerals, resulting in the release of the environmental disruptive chemicals into the soils and creating the "mine soils." Typical mine soils often refer to the antecedent or original soils which are affected and degraded by the acid drainage and mining wastes. Practically, this type of soils also include the exposed parent materials due to accelerated soil erosion and/or the top soil removal for open pit mining, and the deposition of mining solid wastes. Although the properties vary from location to location depending on the local geochemistry, a mine soil is usually acidic, heavy-metal laden, nutrient depleted, highly compacted, and not favorable to plant growth [4].

The strategy for mine soil reclamation is to minimize the environmental impacts of mining by restoring the mine soils and the local ecosystems to the antecedent levels. An adequate reclamation of mine soils not only benefits the local environment but also can contribute to improving the global environment through carbon (C) storage in biomass and in the soils, and thus off-setting the increase of CO_2 emissions from industrial activities. The depleted and drastically disturbed mine soils have a larger potential of C storage over agricultural soils due to the fact that intensively cultivated soils contain relatively high soil organic matter (SOM) and further increasing the C sink capacity is difficult. In contrast, mine soils usually contain low soil organic C (SOC) and thus possess high C storage potentials. Taking full advantages of the available C sink capacity by growing vegetation at the abandoned mining sites would increase the atmospheric CO_2 adsorption and enhance the terrestrial C sequestration [5–7].

Reclamation of mine soils for C sequestration requires high quality of remediation techniques and treatments. It is not enough just to protect against acid mine drainage (AMD) and heavy metals from contaminating the ground and surface waters. The mine soils must be reclaimed in situ so that the grasses, crops, or trees could grow sustainably with limited management. On a long run, a quick establishment of the vegetation and enhancing microbial activities in a mine soil can improve soil quality by accentuating phytoremediation of the contaminants, decreasing soil erosion, and enhancing concentrations of SOM and plant nutrients.

Mine Soil Quality and Amendment Materials

Harnessing an effective CO_2 sink of a mine soil site requires establishment and maintenance of a healthy forest or other vegetation cover for a time scale of at least 25 years. Soils with high quality are indispensable to support the vegetation that can thrive and sustain itself. However, mine soils, especially the gob piles or mining rock wastes, usually have poor soil quality such as low SOM content, low fertility, micronutrient imbalance or toxicity, low N and P availability, soil compaction caused by the grading operations, shallow soil depth, low moisture holding capacity, high electrical conductivity, high heavy metal contents, and extreme pH, which all adversely affect vegetation establishments and SOC sequestration [5]. Therefore, soil amendments and proper managemnet are needed in order to improve the soil physical, chemical, and biological properties at a disturbed site for establishing vegetation and making it an effective atmospheric CO_2 sink. Several natural minerals and agricultural, industrial and municipal wastes have been tried for these purposes as soil amendments. For example, manures [8, 9], composts [10], biosolids [9, 11, 12], and paper mill sludge [9] have been successfully applied to increase the SOM content in the mine soils. Limestones, zeolites [13], and coal combustion byproducts [14–16] (e.g., fly ash, bottom ash, and flue gas desulfurization (FGD) gypsum) have also been intensively researched in reducing mine soil acidity and decreasing the heavy metal toxicity and uptake by plants. A range of various commercial N, P, and K fertilizers have been used to provide adequate nutrients for the vegetation establishment at mining sites as well [8, 12, 17]. The land applications of these conventional amendment materials are also encouraged by the increasing demands on disposal and reuse of these industrial by-products and community wastes at low cost. However, various levels of heavy metals (e.g., Hg, Cd, Cr, and Pb) and other toxic elements (B, As, Se, and Mo) often occur in coal combustion by-products [18]. Nuisance odors, the potential of pathogen transmission, and presence of toxic and persistent organic chemicals and metals in biosolids have for the most part limited the use of land applications [19]. A survey study [20] on 9 different biosolids produced by municipal wastewater treatment plants in 7 USA states indicated that some biosolids were highly enriched in organic wastewater contaminants (OWCs), suggesting the land application of the solids might become a potential nonpoint source of OWCs into the environment. The OWCs included pharmaceuticals, hormones, detergent metabolites, fragrances, plasticizers, and pesticides. Therefore, new types of effective and environmentally safe soil amendment materials are urgently needed for mine soil reclamation.

Using Nanoenhanced Materials as Soil Amendments

Nanotechnology is an advanced modern approach. It provides new types of materials which offer the unique and important solutions to the limitations of other conventional materials and have numerous applications [21]. Nanomaterials and nanostructures have nanoscaled dimensions that range from 1 to 100 nm and often exhibit novel and significantly changed physical, chemical, and biological properties as a result of their structure, larger specific surface area, and quantum effects that occur at the nanoscale [21]. Applications of nanotechnology in water treatment and purification have witnessed significant developments in recent years [22–24]. However, little progress has been made regarding the application of nanoparticles to improve agricultural soil quality and to reclaim the drastically disturbed lands. Lal [25] proposed that applying nanotechnology in agricultural sector was one of the available options to increase the agricultural production, solve environmental problems, and feed the world's growing population. Hence, it is imperative to review the state of the science of nanotechnology that has potentials in mine soil reclamation and mine soil quality improvement and to explore the feasibility of using nanoenhanced materials as replacements for the conventional amendment materials in agriculture. The specific nanotechnology interested in this paper encompass those able to increase soil pH and fertility, improve soil physical structures, reduce mobility, availability, and toxicity of heavy metals and other environmental contaminants and those able to stabilize the soil components and abate soil erosion at a mining site. Therefore, the overall objectives of this paper are to (a) review the available literature on various environmentally-friendly nanoenhanced materials which could be used as in situ soil amendments for mine soil reclamation; (b) briefly discuss the transport and mobility of those nanoparticles in the environment as well as their possible ecotoxicological effects (if any); (c) propose a practical approach to application of the nanomaterials in mine soil reclamation at low cost and in a more environmentally friendly fashion.

NANOMATERIALS FOR SOIL RECLAMATION AND ENVIRONMENTAL REMEDIATION

Reclamation of mine soils involves removing soil contaminants and enhancing soil quality and fertility. Nanotechnology is a promising approach for these purposes. Two advantages of nanomaterials over the traditional amendments for soil reclamation include the higher reactivity due to smaller particle size and higher specific surface area and the easier delivery of the small-sized particles into the porous media (soils). High reactivity leads to a high efficiency and high rate of soil reclamation, while easy delivery is advantageous for in situ application. These nanomaterials with large potentials for mine soil reclamation include zeolites, zero-valent iron nanoparticles, iron oxides nanoparticles, phosphate-based nanoparticles, iron sulfide nanoparticles, and C nanotubes. With emphasis on their functions in soil quality improvements, transport and mobility of those nanoparticles in the environment as well as their possible ecotoxicological effects are also briefly introduced in this section.

Soil Conditioner-Zeolites

Zeolites are crystalline, hydrated aluminosilicates of alkali (Na^+, or K^+) and alkaline earth cations (Ca^{2+} or Mg^{2+}) characterized by an ability to hydrate/dehydrate reversibly and to exchange some of their constituent cations with aqueous solutions, without a major change in structure [26]. Their unique feature is that the zeolites possess an open, three-dimensional cage-like structure and a vast network of open channels extending throughout. The channels and pores, typically 0.3 to 0.7 nanometers in diameter, impart the mineral large specific area (about $105 \, m^2 g^{-1}$) for ion exchange and for selective capture of specific molecules (e.g., H_2O). Because of these structural features, zeolites generally have low density compared with that of other minerals. Nearly 50 natural species of zeolites have been recognized, and more than 100 species have been synthesized in the laboratory [27]. Clinoptilolite is the most abundant zeolite species in the sedimentary deposits on the earth and also the most mined zeolite minerals in the world [28]. Zeolites can occur in soils but with only less than 5% (by weight) in content, and again clinoptilolite is the major zeolite species in soils [28]. Because of their ion exchange, adsorption, and molecular sieve properties, as well as their geographically widespread abundance, zeolite minerals have generated worldwide interest for use in a broad range of applications. In agricultural industries, zeolites have been used as soil conditioners, slow-release fertilizers, and remediation agents for contaminated soils [13]. As a soil conditioner, literatures showed that zeolite nano materials can improve the mine soil quality by increasing the water holding capacity, increasing the clay-silt fractions, improving nutrient levels, and removing toxins [13].

Reducing Soil Bulk Density and Improving Soil Water Holding Capacity

Natural zeolites have several unique physical properties that make them attractive as additives to improve soil physical properties. For example, bulk density of zeolite minerals can be as low as $0.8\,Mg\,m^{-3}$ due to the porous nature [13]. In comparison, mine soils usually have coarse texture (contributing to higher water infiltration rate and lower water holding capacity) and higher bulk density (hindering root growth). Adding fine-grained zeolites (<0.05 mm) to mine soils can increase the effective silt and clay fractions, increase the water-holding capacity, and lower the bulk density thus benefiting the vegetation establishment. Githinji et al. [29] reported application of zeolite (0.55–0.6 mm) at a 15% ($v\,v^{-1}$) rate to sand (0.31 mm) media decreased the bulk density from 1.67 to $1.56\,Mg\,m^{-3}$ and increased the available water content by 2 times. Wehtje et al. [30] attributed the increased performance of bermudagrass (Cynodon dactylon) in zeolite soil mixtures to its increased water holding capacity relative to the unamended soils, but not to any alteration of the chemical properties by the amendments. Zeolite particle size and the application rate are important parameters in improving soil physical properties. Petrovic [31] observed through laboratory studies that the optimum particle size of clinoptilolite added to golf course sand was between 0.1 to 1 mm in order to maximize benefits of water infiltration, water availability, and aeration. Huang and Petrovic [32] concluded that the water available to plants increased when clinoptilolite particles decreased in size and the amendment dosage increased in a sand medium. They observed that plant-available water in sand amended with 5 and 10% ($g\,g^{-1}$) clinoptilolite with a particle size of >1 mm was near $6\,g\,kg^{-1}$; whereas plant-available water in the same medium amended with 5 and 10% ($g\,g^{-1}$) of <0.047 mm clinoptilolite was approximately 10 and $17\,g\,kg^{-1}$, respectively. Huang and Petrovic [33] also reported that sand-based putting green turf could benefit from a 10% clinoptilolite amendment by increasing shoot-growth rate by 26–60%. Lopez et al. [34] proposed a method to ameliorate the drought problem by adding zeolite to the soil and acting as a wicking (capillary) material to draw the water from a shallow ground water table to plant's root zone, thus reducing the dependence on precipitation or irrigation. Their pilot study showed that the grass survived in the zeolites-packed core structures, while the grass planted in the soil at the same site died. These reports show that zeolites could be an effective soil amendment to increase the water availability in mine soils and ensure the survival of the vegetation at the sites where the soils have poor soil structures, high bulk density, low water holding capacity, and the available water mainly depends on precipitation.

Improving Soil pH and Cation Exchange Capacity

Mine soils are usually acidic and infertile with low cation exchange capacity (CEC), resulting in poor nutrient status for plant growth. In contrast, pure zeolite materials usually have high CEC ranging from 220 to 570 $cmol_c$ kg^{-1} [28]. Adding zeolites to a soil can increase the soil's overall CEC and pH in most cases [13], thus soil's nutrient holding capacity is enhanced. For example, Huang and Petrovic [35] applied 10% ($g\,g^{-1}$) zeolite to a sandy soil, increasing the CEC from 0.08 to 15.59 $cmol_c$ kg^{-1} and the pH from 5.4 to 6.6. After applying clinoptilolite to two types of materials (glacial till and marine clay) at a rate of 25% and 50% ($g\,g^{-1}$), the resulting CEC and pH of the mixtures increased 2.6~3.3 times and from 4.2 to 6.5, respectively [36], reflecting the prominent effect of zeolites on raising soil CEC. It was also reported that adding 0.2~2% zeolites to soils was beneficial to crop seed germination and crop productions [37–39].

Zeolites have alkaline properties with pH around 8, which could increase the pH of acidic solution or soils. The acid neutralization property may arise from the high CEC by which zeolites exchange the solution proton (H^+) with Ca^{2+} ion. But the acid neutralization capacity of zeolites is limited compared with agricultural liming materials. Preliminary studies conducted at the Carbon Management and Sequestration Center, the Ohio State University, USA, showed that applying zeolites at 10% ($g\,g^{-1}$) rate to mine soils increased the pH by 0.5–1 unit [40]. Using the liming materials, pH, however, increased by 2 to 3 units [40]. It is not known whether zeolites can suppress the acid production in mine soils resulting from the oxidation of sulfide minerals. But applying small-grained zeolites may fill the pores in the coarse-textured mine soils and reduce the oxygen diffusion to the underlying sulfide materials. Moreover, by adsorption, zeolites are able to remove gaseous molecules such as H_2S and SO_2 and protect the vegetation from being harmed by these gases generated from the same sulfide-containing materials at a mine site.

There is no report using other nanomaterials for improving soil physical properties.

Nanoenhanced Fertilizers

Zeolites-Enhanced Fertilizers

Mine soils usually lack nitrogen (N) and phosphorus (P), and fertilizers are needed to ensure successful vegetation establishment [41]. However, applying conventional N fertilizers often promotes the growth of noxious weeds, suppressing the growth of crops and tree seedlings [41]. Moreover, applying too much fertilizer may increase nitrates leaching to the ground water and cause ground/surface water contaminations due to the coarse texture of the mine soils and the accelerated soil erosion. Therefore, nitrogen-loaded zeolites have been researched in order to provide a slow release of the nutrients to meet the need of crops while reducing the leaching loss of the fertilizer [13], thereby improving the fertilizer efficacy. Moreover, volatilization of gaseous N (e.g., as NH_3 or N_2) can also be minimized if N H4 +-type fertilizers are exchanged onto zeolite exchange sites so that the N H4 + ion is unavailable for conversion into gaseous phase via microbial processes [13].

Clinoptilolite is highly selective for K^+ and N H4 + relative to sodium (Na^+) or divalent cations such as Ca^{2+} and Mg^{2+} due to the location and density of negative charge in the structure and dimensions of interior channels [13]. Hence, NH_4- and K-loaded zeolites are typically used as slow release fertilizers. For example, Perrin et al. [42] loaded clinoptilolite with N H4 + by soaking the various size fractions in 1 M $(NH_4)_2SO_4$ for 10 days (d), changing the soaking solution every 2 to 3 d, then applied the solid to 4-liter containers seeded with sweet corn (Zea mays). They [42] observed that the soil fertilized with $(NH_4)_2SO_4$ leached 10 to 73% of the added N (depending on applying N rate) whereas mere <5% of the added N leached from the $(NH_4)_2SO_4$-zeolite-amended soil regardless of the N application rate and zeolite particle size. Nitrogen use efficiency (NUE) ranged from 72.0 to 95.2% in N H4 + clinoptilolite-amended soils after 42 d of plant growth, compared with NUE of 29.7 to 76.3% in soils fertilized with $(NH_4)_2SO_4$ only. Moreover, Lewis et al. [43] not only observed that the NH_4-loaded clinoptilolite was an efficient slow-release N-fertilizer but also found that the amendment could prevent injury by urea to radish (Raphanus sativus) plants. Based on these evidences, Barbarick and Pirela [44] proposed that zeolites could be used effectively in agriculture to prevent leaching losses of ammonium-type fertilizers, to reduce ammonia toxicity to plants, and to increase agronomical yields. Potassium-loaded zeolites have also been researched as a slow-release K-fertilizer [45, 46].

Phosphorus (P) is also an important nutrient indispensable for vegetation establishment and reforestation in the reclaimed mining areas. Rock phosphates such as apatites (e.g., $Ca_{10}(PO_4)_6(OH)_2$) are commonly used sources of P in mine soil rehabilitation [17]. But the availability of the phosphorus from the rocks depends on the apatite dissolution in the soil. Alkaline soil pH often impedes the dissolution and decreases the soluble P amount. For example, Jacinthe and Lal [17] observed no effect of rock phosphate on the tree growth in a reclaimed mine land, probably due to the relatively high pH of the soil ranging from 6.5 to 8.0. Zeolites have also been used to alleviate this problem: some researchers have used a mixture of zeolite and ground apatite to increase the dissolution of the latter to provide more available P even at high soil pHs. The idea is to create exchange sites or a "sink" for Ca^{2+} in zeolites. A decrease of the Ca^{2+} in the soil solution by this process promotes further apatite dissolution and phosphate release. This hypothesis was first tested by Lai and Eberl [47] and confirmed by other researchers [48–50]. Lai and Eberl [47] mixed a rock phosphate with untreated and treated (N H4 +, Na^+, and H^+) zeolite at a ratio of 1 : 5 and observed that the mixture contained 5–70 times higher soluble P than contained in rock phosphate-only control. Using batch experiments, Allen et al. [49] showed that the higher zeolite to P rock ratio, the more P was released from the mixtures to the solution, further confirming the role of zeolites in P rock dissolution. These results suggested that zeolites could improve the effectiveness of rock phosphate used as P fertilizer in mine soil reclamation.

Other Nanoenhanced Fertilizers

Except for zeolites-enhanced fertilizers, there are few reports on other type of nanomaterial-related fertilizers. Concerned by the mere 30–50% efficiency of the conventional fertilizers and no other management practices to enhance the rate, Derosa et al. [51] urged to apply nanotechnology to fertilizer developments. Lal [25] also recommended that applying nanotechnology in agriculture (including fertilizer development) is one of the available options to increase the crop production and feed the world's increasing population. The observations that C nanotubes and zinc oxide nanoparticles are able to penetrate tomato (Lycopersicon esculentum) plant root or seed tissues indicate that new nutrient delivery systems can be developed through exploiting the nanoscale porous domains on plant surfaces [51]. Liu and Zhao [52] and Liu [53] discussed nanosized vivianite ($Fe_3(PO_4)_2\cdot8H_2O$) particles (~10 nm) and apatite ($Ca_5(PO_4)_3Cl$) particles (<200 nm) for heavy metal remediation. These phosphate-based nanoparticles have potentials to be used as P nanofertilizers for agricultural uses.

Nanomaterials for Remediating the Mine Soils Contaminated with Heavy Metals and Other Toxins

Zeolites

Natural and synthesized zeolites can immobilize heavy metals and radioactive nuclides in contaminated soils and sediments, thus reducing the risks of those toxins being released to neighboring water bodies or taken up by plants/animals. For instance, Edwards et al. [54] treated mine soils contaminated by Zn, Pb, Cu, and Cd using synthesized zeolites at rates of 0.5–5% by weight. They observed significant reductions (42%–72%) of the labile and easily-available fractions of the heavy metals after the treatments. In addition to adsorption, soil pH increase caused by zeolites also played a role in the metal immobilization [54]. Similar results have been reported by others [55–57] who use different leaching solutions such as 0.01 M $CaCl_2$ or dilute acetate solution to evaluate the stability of the heavy metals in the soil phase. The leachable fraction of the metals by these solutions was significantly reduced after the contaminated soils were amended with 0.5 to 16% zeolites by weight [55–57]. Plants were also used as indicators to evaluate the metal toxicity and bioavailability in the zeolites-amended soils. Using rye grass (Lolium perenne L.) and alfalfa (Medicago sativa L.) as indicator plants, Haidouti [58] observed that application of zeolite at 1–5% ($g\,g^{-1}$) rates reduced plant uptake of Hg from a contaminated soil by up to 58% in the roots and 86% in the shoots. Chlopecka and Adriano [59] found that adding 1.5% ($g\,g^{-1}$) zeolite to a Zn-spike soil was able to ameliorate the detrimental effect of the metal and to enhance the growth and yields of maize and barley (Hordeum vulgare). The Zn concentration in plant tissues was also reduced by the amendment. Knox et al. [60] reported that applying 2.5–5% zeolites to a metal-laden soil near a Zn-Pb smelter substantially enhanced the growth of maize and oat (Avena sativa) and decreased the Cd, Pb, and Zn accumulations in the plant tissues. In contrast, neither plant could grow in the unamended soil. Mahmoodabadi [61] indicated that application of natural zeolites increased the shoot dry weight, the number and dry weight of the root nodule and abated the Pb toxicity to the soybean (Glycine max). However, there are also quite a few reports which indicated that application of zeolites reduced the growth of some crops and vegetables [62–64]. It is generally believed that use of Na-type zeolites resulted in release of Na^+ to the soil solution and negatively affected the plant growth even though the adverse effects of the heavy metals were alleviated. Therefore, using Ca-type zeolites for heavy metal remediation is preferred at the sites where revegetation is planned.

Additionally, possessing unique selectivity for Cs^+ and Sr^{2+}, zeolites are also good remediation agents for trapping radioactive ^{139}Cs and ^{90}Sr from contaminated soils due to nuclear fallout, contact with water from reactor cooling reservoirs, or radioactive waste spills [13]. Similar to heavy metal remediation, the primary purpose of using natural zeolites is to immobilize radionuclides in the soil and to reduce or prevent the uptake of those by plants [13].

Iron Oxides Nanoparticles (nFeOs)

As an important constituent of soil and an essential nutrient to plants and animals, iron (Fe) is ranked the 4th most abundant element in the earth. The Fe oxides found in soils and sediments usually occur as nanocrystals (5–100 nm in diameter) with reactive surfaces capable of sorbing a wide range of both inorganic and organic species through mechanisms such as surface complexation/surface precipitation [65]. Due to their prominent absorption capacity for toxins and their environmentally friendly characteristics, a variety of engineered iron oxide nanoparticles have been fabricated and applied to in situ water/ soil remediation processes. For example, nano-Fe oxides (nFeOs) solution can be pumped/spread directly to contaminated sites at low cost with negligible risks of secondary contamination. The intensively studied nFeOs for heavy metals removal from water/wastewater include goethite (α-FeOOH, needle-like, 200 nm × 50 nm), hematite (α-Fe$_2$O$_3$, granular, 75 nm), amorphous hydrous Fe oxides (particles, 3.8 nm), maghemite (γ-Fe$_2$O$_3$, particle, 10 nm), and magnetite (Fe$_3$O$_4$, particles, around 10 nm) [66]. Those nFeOs have been widely researched for heavy metal removal form aqueous phase through adsorption. The target contaminants included Cu^{2+}, Cr^{6+}, Ni^{2+}, Pb^{2+}, Cr^{3+}, Zn^{2+}, As^{+5}, and As^{+3} [66]. However, the use of nFeOs for contaminated soil reclamation has not been widely studied. But the capacity of the nanoparticles for removal of heavy metals from aqueous phase suggests that these particles are able to sequester the labile fractions of heavy metals from the soil solution through adsorption and thus reduce the availability and mobility of those toxins in the soils. Moreover, addition of industrial wastes rich in iron oxides to contaminated soils has been a common practice for heavy metal immobilization [67–69], suggesting that mixing nFeOs with the mine soils could effectively immobilize the soil bound toxic metals. Shipley et al. [70] applied As-spiked solution to a column packed with soil mixed with 15% (g g^{-1}) nanomagnetite and observed that negligible As concentrations occurred in the effluent for up to 132 days as the influent containing 100 μg L^{-1}. As injected through the column at a rate of 0.3 mL h^{-1}. Only 20% of the contaminant leached out after 208 days. A subsequent batch test suggested that the soil alone had no adsorption of As. Shipley and colleagues [70] also reported that As and other 12 heavy metals (V, Cr. Co, Mn, Se, Mo, Cd, Pb, Sb, Tl, Th, and U) could be simultaneously removed by the nFeOs in the soil. After 35 hours of the leaching test, only Cr, Mo, Sb, and Co leaching reached more than 20% of the influent levels, revealing the fairly strong and high adsorption capacity of the nanoparticles even for multiple contaminants. Nanohematite has an adsorption capacity similar to the nanomagnetite [70].

Besides the chemical compositions, remediation efficiency and deliverability of the nanoparticles are largely controlled by their stability and transport behaviors in the media (water, soil, or aquifer). Stability and transport of nFeOs depend on the particle size, particle concentration, particle magnetism, the solution chemistry, and the medium property. For a given nanoparticle suspension, the particle stability is generally governed by the electrostatic repulsion between particles [71]. The force is generated by the particle surface charge. and surface "zeta potential" is used to quantify the magnitude of the charge or the electrostatic repulsion. The higher the zeta potential is, the stronger the repulsion force would be between particles, thus the more stable the nanosolution is. Charged ions (e.g., H^+, OH^-, Na^+, or Cl^-) in the background solution can affect the suspension stability by changing the particles surface charge (zeta potential). A pH value where the net surface charge becomes zero is called "point of zero charge" (PZC), and the solution is least stable and most prone to form aggregates at pHs close to the PZC. Therefore, the influences of a solution pH on the nanoparticles stability depend on how close the solution pH is to the particle PZC. For example, the PZC is at pH 7.1 for magnetite nanoparticles. The suspension was not stable at pHs from 6 to 8 because the net particle surface change reduced to around zero and fast aggregation took place due to the minimum repulsion. In contrast, the nanoparticles solutions remained stable at pHs from 3 to 5 or from 9 to 10, which were far from the PZC of magnetite nanoparticles [72]. In these cases, the average particle size remained similar to the original size (60 nm) [72].

Nanoparticles in a concentrated solution more likely collide with each other and form aggregates and precipitates than in a dilute solution, so the former solution is less stable than the latter [73, 74]. He et al. [73] reported that aggregation rates were higher for smaller hematite nanoparticles due to changes of the surface properties with particles size changes. More importantly, for nFeOs with strong magnetism, the additional attractive force of magnetism among the particles increases the probability of aggregation. In other words, the stability and transport of magnetic nanoparticles are negatively influenced by a combination of electrostatic and magnetic interactions as observed by Hong et al. [75]. Through column test with sand media, they [75] reported that the less-magnetic nanoparticles eluted from the columns more than the more-magnetic particles. And the nonmagnetic nFeOs were transported the most. The majority of particles were retained at the column inlet for all transport experiments, with the greatest amount of retention being that of the magnetic nanoparticles, indicating that magnetically induced aggregation and subsequent straining resulted in greater retention in the column. Magnetic particles include maghemite (γ-Fe_2O_3), magnetite (Fe_3O_4), and zero valent iron (Fe^0), while hematite (α-Fe_2O_3) nanoparticles are nonmagnetic. On the other hand, transport of those magnetic nanoparticles might be controlled by the imposing of an external magnetic field to the system.

Natural organic matter is able to modify the nanoparticles surface and change the particle PZC when absorbed by the latter. Therefore, changes of a nanoparticle suspension stability by humic acids (HA) could be explained with how the acids affect the particle PZC. Adsorption of HA often results in a decrease of magnetite PZC towards the more acidic pH values, and the more HA is added to the solution, the lower PZC becomes. For example, Hu et al. [72] reported PZC of magnetite nanoparticles dropped from 7.1 (without HA) to 5.8 at 2 mg L^{-1} HA and to 3.77 at 3 mg L^{-1} HA. When the HA concentration was high enough (e.g., 10 mg L^{-1}), the PZC was dropped to pH values out of the range (pH 3–10) that is commonly encountered by the natural environment. In this case, the suspension shows the highest stability under normal conditions [72]. Similar results are also reported by others [73, 76]. In addition, an increase of the solution ionic strength generally enhances the aggregation of the nanoparticles [72].

Iron oxides nanoparticles are generally believed to have low or no toxicity to the living organisms according to limited related reports. For example, Karlsson et al. [77] evaluated the ability of the metal oxides particles with varying sizes to cause cell death, mitochondrial damage, DNA damage, and oxidative DNA lesions after exposure of the human cell line A549. They concluded that the iron oxide (Fe_2O_3) nanoparticles showed low toxicity and no clear difference between the different particle sizes. Auffan et al. [78] believed the organic coating on maghemite nanoparticles served as a barrier for a direct contact between particles and the cells (human fibroblasts), further reducing the possible toxic effects. They found that the coated nFeOs produced weak cytotoxic and no genotoxic effects. One main mechanism behind the toxicity of manufactured metal nanoparticles is their ability to cause oxidative stress in cells by producing reactive oxygen species (ROS). ROS can damage proteins, lipids and DNA as well as give rise to necrosis and apoptosis [77]. However, Limbach et al. [79] believed that the chemical composition rather than the nanoscale size is a most decisive factor determining the formation of ROS in exposed cells. Moreover, they observed that dissolved iron ions promote a 20 times higher ROS production than exposure to the same amount of iron in the form of Fe_2O_3 nanoparticles, suggesting nanosized iron particles do not exert more toxicity than the soluble irons or solid irons with larger particle sizes. As a matter of fact, Sadeghiani et al. [80] suggested that polyaspartic-acid-coated magnetite nanoparticles may be considered as a potential precursor of anticancer drugs.

Nanoscale Zero-Valent Iron Particles (nZVI)

Nanoscale zero-valent iron (nZVI) technology developed in 1990s was initially designed to destroy the toxic halogenated hydrocarbon compounds and other petroleum-related products which entered the ground water environment through gas tank leakage, organic solvent spills, etc. [81]. The metallic iron particles are highly effective reducing agents and able to convert several persistent organic contaminants to benign compounds by reduction reactions. These contaminants include chlorinated methanes, chlorinated benzenes, pesticides, polychlorinated biphenyls (PCBs), and nitroaromatic compounds [81]. In addition to the high decontamination effectiveness, this technology possesses the advantages of using an environmentally friendly material and being easily delivered to the subsurface environment due to the small particle size.

This technology is also used to treat heavy metals in water and soil. Zero valent iron is a strong reductant with a reduction potential ($E0$, Fe^{2+}/Fe^0) of -0.44 V [71]. Theoretically, some metals with $E0$ much more positive than -0.44 V could be reductively immobilized by nZVI. Typical examples of such metals with environmental importance include $CrO_4\,2\,-/Cr^{3+}$ ($E0 = +1.56$ V), $Cr_2O_7\,2\,-/Cr^{3+}$ ($E0 = +1.36$ V), and $UO_2\,2\,+/U^{4+}$ ($E0 = +0.27$ V) [71]. The high-valent species ($CrO_4\,2\,-$, $Cr_2O_7\,2\,-$, and $UO_2\,2\,+$) of those metals are usually more soluble and more toxic than their low valent counterparts (Cr^{3+} and U^{4+}) in the natural environment. nZVI is able to transform the former to the latter through reduction reactions, thus decreasing the solubility/mobility and toxicity of those metals (the whole process is called reductive immobilization). For example, uranium (U) is the most common radionuclide contaminant found at many nuclear waste sites. It is mainly detected in contaminated groundwater as highly soluble and mobile U^{6+} in the form of $U O2\ 2+$ [82]. In soils and in uranium mining tailings, $U O2\ 2+$ sorbs onto Fe oxyhydroxides [83]. However, acid mine drainage can dissolve and release the sorbed uranium to the nearby ecosystem. These risks can be remediated by reducing it to insoluble U^{4+} oxides by nZVI. Several related reports have shown that, compared to the other reductants such as metal iron filing, galena (PbS), and iron sulfide, nZVI is very efficient to reductively immobilize U^{6+} from aqueous phase, which could be attributed to its nanosize, high reactivity, large surface area, and reactive Fe(II) produced by nZVI [84–88]. This literature confirmed that U^{6+} was predominantly removed by nZVI via reductive precipitation of $U O2\ 2+$ (U^{4+}) with minor precipitation of $UO_3 \cdot 2H_2O$ (U^{6+}) as confirmed by the X-ray photoelectron spectroscopy (XPS) and X-ray diffraction (XRD) analyses. Oxygen level, solution pH, and presences of bicarbonates and calcium ions all affect the reductive immobilization processes [84, 85]. It has also been reported that nZVI was able to reduce higher valent Cr^{+6} to low valent Cr^{+3} in water or soil media. Franco et al. [89] reported that 97.5% of Cr^{+6} in a contaminated soil could be reduced to Cr^{+3} by nZVI, which significantly reduced the chromium toxicity in the spoil. Similar reductive immobilization of Cr^{+6} in soils by nZVI was reported by others [90, 91].

Selenium (Se) is an essential nutrient in animal systems, but high concentrations can threaten biological systems when human activities, such as mining into shale for oil and phosphorus or irrigating arid and semiaridlands, produce seleniferous soils [92]. Plants can accumulate Se from the impacted soils [93]. Plant accumulation and soil ingestion lead to Se bioaccumulation and Se poison in livestock and wildlife [94, 95]. Similar to uranium and chromium, high-valent selenium species (S e O4 − 2 or Se^{6+} and S e O3 − 2 or Se^{4+}) are more soluble and mobile in the natural environment and more toxic than the low-valent species such as Se^0 and Se^{-2}. nZVI has been applied to remove the selenium from the solution and reduce the high-valent species to the low-valent ones thus the toxicity and solubility of Se are greatly lowered [71]. Olegario et al. [96] reported that nZVI had high uptake capacity for removal of dissolved Se^{6+} up to 0.1 mole Se/mole Fe. Using X-ray absorption near edge structure (XANES) spectroscopy and X-ray absorption fine structure (EXAFS) spectroscopy, they identified FeSe compound in the solid phase as the reduced Se^{2-} species transformed from S^{6+}. They concluded that nZVI was capable of an efficient reduction of soluble Se oxyanions to insoluble Se^{-2}.

nZVI is also able to treat some other toxic elements in water or soil such as Hg^{+2}, Ni^{+2}, Ag^{+1}, Cd^{+2}, As^{+3}, and As^{+5} [97–101]. The decontamination mechanisms include reduction of metal ions to zero valent metals on the nZVI surfaces and/or adsorption of the ions on the nZVI particle shells which consist of a layer of iron oxidation products (iron oxides) [71]. For example, Watanabe et al. [102] reported that applying 0.01% nZVI ($g\,g^{-1}$) to a Cd-spiked soil significantly reduced the Cd accumulations in rice (Oryza sativa) seeds and leaves by less than 10% and 20% of those without nZVI amendment.

The environmental migration of bare nZVI has been estimated to be within a few centimeters under subsurface environment [103, 104] due to quick nanoparticles agglomeration and interactions with surfaces of the ambient porous media. Substantial efforts have been made to increase the stability and mobility of nZVI (e.g., using nanoparticle stabilizers), hoping that nZVI disperses the entire contaminated aquifer and degrades the pollutants in situ as soon as being injected underground through one or more injection wells. Supported by the laboratory column test results, quite a few reports have claimed successful synthesis of nZVI with improved stability and mobility as well as reactivity [105–108]. But there is no solid evidence on significantly increased mobility of such products in the field [71]. Stabilized nZVI has been visually confirmed to travel merely 1 m from an injection well, and evidence suggests that the maximum travel distance of up to 2-3 m may be achieved in high permeability formations [71]. The discrepancies between the lab reports and the field tests resulted from the fact that laboratory applied lower Fe concentrations (<0.25 g/L), higher flow velocities (15–30 m/day), and simplified subsurface simulations by sand-packed columns. As a matter of fact, much higher Fe application rates (1–30 g/L), lower groundwater flow rates (0.1 to 10 m/day), and much more complicated aquifer formations were found under the field conditions [71], which favored aggregation and precipitation of nZVI. In addition, dissolved oxygen very rapidly oxidizes nZVI, forming maghemite and magnetite precipitates [109]. These facts suggest that risks of nZVI spills in the environment and subsequent exposure of organisms to the nZVI are not significant on the current stage of nZVI technology.

There is no field study on applying nanoparticles for soil remediation. However, there are some differences from groundwater remediation. For mine soil reclamation and vegetation establishment purposes, a thin soil surface layer (e.g., 50 cm deep) for plant root establishments is usually interested. Ideally, the nanoparticle suspension would be applied to all over the targeted land surfaces in a way similar to the surface irrigation. By manipulating the nanoparticle size, the particles would be ideally retained within the contaminated surface layer only after the whole targeted soil column was saturated and treated by the particles, thus reducing the risks of nanomaterials spill and avoiding secondary contaminations to the neighboring water bodies. From this point of view, nZVI and other nanoparticles with extremely high mobility are not needed for surface soil remediation purpose.

There is a limited number of peer-reviewed and published studies pertaining to the toxicological and ecotoxicological effects of nZVI application in the environment [110]. Based on available information, Grieger et al. [110] summarized the possible effects of exposure to nZVI as follows: (a) acute toxicity to aquatic organisms appears to be relatively low, and sublethal effects have been observed at lower concentrations (<1 mg L^{-1}); (b) nZVI can attach to organisms and cells and cause histological changes and morphological alterations in some species; (c) some coatings have also been found to decrease toxicity mainly through reduced adherence; (d) effects are thought to be linked to the release of Fe(II) from nZVI and subsequent ROS production as well as disruption of cell membranes leading to cell death and lysis and possible enhancement of biocidal effects of Fe(II); (e) the aging of nZVI under aerobic conditions reduces nZVI toxicity, whereby Fe^0 is rapidly oxidized

Other metal-based nanoparticles for environmental remediation encompass nanoscale manganese oxides and hydroxides, aluminum oxides, titanium oxides, zinc oxides, and magnesium oxides. All of those nanoparticles could remove heavy metal from solution by surface adsorption, a mechanism similar to that of heavy metal removal by iron oxides [65]. Among those metal oxides nanoparticles, iron and manganese nanoparticles are sensitive to the reduced environment such as those in a waterlogged soils or wetlands. Those particles may be reduced to the lower valent states and lose the adsorption capacity. For manganese, zinc, and aluminum-based nanoparticles, phytotoxicity might occur if those particles are applied to the acidic soils. Moreover,

Limbachet al. [79] reported that cobalt and manganese oxides (Co_3O_4 and Mn_3O_4) nanoparticles produced more ROS (indicating more toxicity) than their respective salt solutions while titanium oxide (TiO_2) and iron oxide (Fe_2O_3) nanoparticles were relatively inert.

Phosphate-Based Nanoparticles

Different from nFeOs or nZVI, phosphate-based nanoparticles remediate the heavy metal-contaminated soils by forming highly insoluble and stable phosphate compounds. A typical example is treatment of the lead-laden soils. The solubility products of common lead compounds in soils such as anglesite ($PbSO_4$), cerussite ($PbCO_3$), galena (PbS), and litharge (PbO) have been measured as $10^{-7.7}$, $10^{-12.8}$, $10^{-27.5}$, and $10^{+12.9}$, respectively [111]. In comparison, lead phosphate compounds such as pyromorphites (($Pb_5(PO_4)_3X$, X = F^-, Cl^-, Br^-, and OH^-) have solubility products less than 10^{-71} [111]. This fact indicates that lead phosphates are considerably less soluble than other Pb phases generally observed in soils. A transformation of the less stable Pb species to more stable species by phosphate amendments is a thermodynamically favored process which spontaneously decreases the leachability and availability of the lead in the solid phase. Some phosphate amendments have been the most effective method for in situ lead immobilization and have been intensively studied [111]. Other metals having been investigated and reported effective include Cu^{2+}, Zn^{2+}, Cd^{2+}, Co^{2+}, Cr^{3+}, Ba^{2+}, U^{6+}, and Eu^{3+} [112–115]. Generally, soluble phosphate salts and particulate phosphate minerals are commonly utilized forms of the phosphates for this purpose. The former includes phosphoric acid [116], NaH_2PO_4 [117], and $(NH_4)_2HPO_4$ [115], the latter involves various forms of apatite including synthetic apatites [118], natural rock phosphates [112, 115], and biogenic apatites such as fishbone [119]. Although both are highly effective for in situ stabilization of heavy metals at the laboratory scale, problems still exist in the application of these materials in the field. For instance, soluble phosphates, although highly mobile in the subsurface and thus more effective in heavy metal stabilization, may cause the secondary environmental problems of eutrophication. Furthermore, application of phosphoric acids and ammonium phosphates in large amounts may result in acidifying the soils [115]. Amendment dosage of 3% PO_4 (or 1% as P) by weight for soils has been proposed and practiced by USEPA and other government agencies [120], suggesting higher possibility of the phosphate spill to water bodies and soil acidification following the heavy metal remediation.

Yet, the application of solid phosphate is hindered by the large size of the particles, which limits the phosphate mobility and delivery and prevents phosphate from reaching and reacting with heavy metals in subsoil layers. Even the finely ground solid phosphate particles are not mobile in soils, and mechanical mixing is usually needed but not practical in the field for in situ treatment processes. In light of these problems related to phosphates application, Liu and Zhao [52] synthesized nanosized iron phosphate particles for heavy metal immobilization as the commonly used phosphates while overcoming the delivery problem and secondary contamination risks associated with the latter. For example, the nanoparticle suspension, which possesses the same mobility as aqueous solution due to the nanoscaled particle size, is easily delivered to the contamination site with conventional engineering methods (e.g., spray or well-injection). The nanoparticles are also environmentally sound because the phosphate in solid form is much less bioavailable to the algae than those in soluble forms [121]. Algae-bioavailable P and N are primarily responsible for eutrophication in surface waters.

Liu and Zhao [52] prepared and tested a new class of iron phosphate (vivianite) nanoparticles for in situ immobilization of Pb^{2+} in soils. Batch test results showed that the nanoparticles could effectively reduce the leachability and bioaccessibility of Pb^{2+} in three representative soils (calcareous, neutral, and acidic), evaluated by the toxicity characteristic leaching procedure (TCLP) and physiologically based extraction test (PBET), respectively. When the soils were treated for 56 d at a dosage ranging from 0.61 to 3.0 mg g^{-1}-soil as $PO4 - 3$, the TCLP leachable Pb^{2+} was reduced by 85–95%, and the bioaccessible fraction was lowered by 31–47%. Results from a sequential extraction procedure showed a 33–93% decrease in exchangeable Pb^{2+} and carbonate-bound fractions, and an increase in residual-Pb^{2+} fraction when Pb^{2+}-spiked soils were amended with the nanoparticles. Addition of chloride in the treatment further decreased the TCLP-leachable Pb^{2+} in soils, suggesting the formation of chloropyromorphite minerals. Compared to soluble phosphate used forin situ metal immobilization, application of the iron phosphate nanoparticles resulted in around 50% reduction in phosphate leaching into the environment. Liu [53] also reported an effective remediation of a lead-laden soil from a shoot range using synthesized apatite nanoparticles. Laboratory tests exhibited that the apatite nanoparticles solution could effectively reduce the TCLP-leachable Pb fraction in the Pb-contaminated soil from 66.43% to 9.56% after one-month amendment at a ratio of 2 mL solution to 1 g soil and the resulting Pb content in the TCLP solution was reduced to 12.15 mg L^{-1} from 94.33 mg L^{-1}. When the amendment ratio was increased by 5 times, the leachable Pb was reduced to 3.75 mg L^{-1} with only about 3% of the soil Pb leachable. The original soil sample contained an average of 2647.9 mg Pb kg^{-1} soil [53].

These phosphate-based nanoparticles also have potentials to be used as P nanofertilizers. In addition to providing nutrient P to the plants, these nanoparticles also have the advantage of easy delivery (by spraying to the soil surface) with least P leaching to the neighboring water bodies.

Iron Sulfide Nanoparticles

Similar to the mechanisms of heavy metal immobilization by the phosphate-based nanoparticles, sulfide-based nanoparticles have been researched specifically to eliminate the contaminations of mercury (Hg) and arsenic (As) in water and soil/sediment by providing sulfide (S^{-2}) ligands and/or coordination surfaces. As a matter of fact, reduced sulfur (S^{2-}) has been regarded as a stabilizer/sink of heavy metals in the reduced environment such as in the sediments or water-logged soils by forming highly insoluble metal sulfides [122]. It has been proposed that a sediment sample would be regarded safe or non-toxic to the aquatic organisms if the molar ratio of the acid volatile sulfide (AVS) to the total heavy metal concentrations (e.g., Cu + Ni + Zn) was greater than 1 [123]. In this case, theoretically, the heavy metals are all bound in the insoluble metal-sulfide phases and thus the soluble (bioavailable) metals in the pore water are minimized [123]. Moreover, sulfide (S^{2-}) has been widely believed as the most important inorganic ligand to remove the Hg from the water column and suppress the formation of the notorious methyl-mercury (CH_3Hg) in the natural environment. Methyl-mercury has been believed as is one of the most toxic Hg species which can easily bioaccumulate in fish and other aquatic organisms and become biomagnified through food webs. Consumption of fish and shellfish contaminated with CH_3Hg is the primary route of human exposure to mercury [124]. Dissolved, neutral mercury complexes (primarily HS^0 and $Hg(HS)_2 0$) rather than Hg^{2+} or total dissolved Hg are considered the main Hg(II) species controlling the extent of mercury methylation in the contaminated sediments [125, 126]. Iron sulfide amendments can effectively decrease the concentrations of the neutral mercury complexes by formation of charged Hg(II)-polysulfides (e.g., HgS_2^{2-}, $HgSH^+$, HgS_2H^-) [124, 127]. In addition, formation of the insoluble mercuric sulfide complexes also reduces conversion of the ionic Hg to volatile metal Hg in soil [128]. Liu et al. [124] reported that synthesized mackinawite (FeS) was able to remove the aqueous Hg around $0.75\, mol\, Hg^{2+}$/mole FeS. They believed that 77% of Hg removed was through precipitation by forming in soluble HgS species and the remaining 23% was removed by adsorption on the FeS surface. Meanwhile, under anoxic environments, iron sulfides are also able to reduce the mobility and availability of toxic element As by adsorption and/or precipitation processes, depending on the solution pH and iron sulfide type and oxidation state of As [129–132]. For example, Wolthers et al. [130] reported that the maximum As(V) adsorption by FeS occurred at pH 7.4 with an adsorption capacity of 0.044 mol As/mol FeS while the capacity was 0.012 As/mol FeS to As(III) but less pH dependent. Furthermore, the reduction capacity of iron sulfides is also applied to reductive immobilization of Tc^{+6} [133], Cr^{+6} [134], and U^{+6} [135], and reductive degradation of trichloroethylene (TCE) and tetrachloroethylene (PCE) [136–138]. Again, sulfide ion (S^{2-}) plays major role in those reduction reactions, and the decontamination mechanisms are similar to those of zero-valent iron nanoparticles as discussed in Section 2.3.3.

Mackinawite is a widely reported iron sulfide synthesized for those environmental remediation studies in the laboratory. This compound is prepared by simply mixing Fe^{2+}-containing and S^{2-}-containing salts together under anaerobic condition. This method produces black-colored micrometer-sized particles [121, 127, 133], which aggregate and precipitate in a few minutes [127]. Using carboxymethylcellulose (CMC) as nanoparticle stabilizer, Xiong et al. [127] synthesized stable FeS spherical nanoparticle suspension which reportedly remained suspended for at least 3 months with the final average particle size of $31.4 \pm 4\, nm$. Dentrimer was also used as a stabilizer to prepare FeS nanoparticles by Shi et al. [139], forming spherical-shaped particles with 4–6 nm diameter. Xiong et al. [127] showed that the CMC-stabilized nanoparticles were highly effective to immobilize Hg in a sediment sample. For instance, when the FeS-to-Hg (sediment-bound) molar ratio was increased to 26.5, the Hg concentration in the sediment pore water was reduced by 97% and the TCLP leachability of the sediment-bound Hg was reduced by 99%, suggesting the FeS nanoparticles amendment greatly reduced the labile Hg portion in the sample. Hg speciation modeling in their study also indicated that the FeS amendment greatly reduced the concentration of the bioavailable Hg species ($HgS^0 + Hg(HS)_2 0$) by up to three orders of magnitude. Most importantly, the stabilized FeS suspension was highly mobile in a clay loam sediment column, reflecting the intrinsic properties of the nanoparticles and the high deliverability for soil/sediment remediation. They reported that complete breakthrough of the nanoparticles occurred at around 18 pore volumes (PVs), compared to 3 PVs for the inert tracer (Br^-). In contrast, when nonstabilized FeS particles were subjected to the same tests, nearly all (>99.7%) the particles were intercepted on top of the sediment column [127]. Xiong et al.'s work is probably the only one using real FeS nanoparticles to remediate the soil-bound contaminants (Hg). But literatures cited earlier in this section suggest that FeS nanoparticles would be excellent candidates for in situ immobilization of other heavy metals (especially As) and some organic pollutants bound in soils or in sediments.

However, cautions must be taken when a mine soil reclamation plan is proposed to use FeS: first of all, most of the iron sulfide (S^-) solids are not stable under the aerobic environments and are easily oxidized to soluble sulfate species (SO_4^{2-}) by the air [122, 133], thereby their adsorption capacity is lost and the contaminants already retained on the FeS solid surface would be rereleased to the pore water and become remobilized [122]. Processes such as draining a pond or a water-logged land and dredging the sediments are a few examples of exposing the sediments to the air. Practically, it is difficult to maintain a soil/sediment under anaerobic environment for long period, and a change of the redox potential might cause a secondary contamination problem related to FeS amendments. Secondly, acid mine drainage (AMD) is one of the serious environmental concerns at most of the abandoned mining sites. Huge efforts have been made on research, prevention, management, and remediation of AMD and acidic mine soils for many decades [140]. As a matter of fact, the acidity in the drainage and in the soils originates from oxidation of the iron sulfide minerals (mostly pyrite, FeS_2) by oxygen (O_2) after these buried minerals were exposed to the air through the mining operations [140]. Therefore, simply adding the FeS minerals to the soils as suggested by the literatures cited above might exacerbate the AMD and soil acidity problems at a mining site. More stable immobilization agents such as iron oxide nanoparticles (for As) or phosphate-based nanoparticles (for heavy metals) should be better options.

Carbon Nanotubes

The C nanotubes (CNTs) are C macromolecules consisting of sheets of C atoms covalently bonded in hexagonal lattices that seamlessly toll into a hollow, cylindrical shape with both ends normally caped by fullerene-like tips [141]. Based on the structures, CNTs are categorized into two main classes: single-walled C nanotubes (SWCNT) and multiwalled C nanotubes (MWCNT). The lengths of CNTs can range from several hundred nanometers to several micrometers, and the diameters from 0.2 to 2 nm for SWCNT and from 2 to 100 nm for coaxial MWCNT. The large surface area, tubular structure, and nonpolar property make CNTs a promising adsorbent material for nonpolar organic contaminants in an environmental media, such as trihalomethanes, polycyclic aromatic hydrocarbons, or naphthalene, dioxin, herbicides, DDT and its metabolites [22, 23, 141]. Compared to an activated C, the purified CNTs possess two to three times higher adsorption capacities for organic contaminants [22].

Due to the nonpolar property of the C material, sorption of the polar metal ions by raw CNTs is very low but significantly increases after the CNTs surface is chemically modified and a large amount of oxygen-containing polar functional groups (–COOH, –OH, or –C=O) are created. These functional groups cause a rise in negative charge on C surface, and the oxygen atoms in functional groups donate single pair of electrons to metal ions, consequently increasing the cation adsorption capacity of CNTs [142]. For example, MWCNTs, pretreated with nitric acid, have been used successfully for the sorption of different heavy metal ions, including Pb(II) ($97.08 \, mg \, g^{-1}$), Cu(II) ($24.49 \, mg \, g^{-1}$), and Cd(II) ($10.86 \, mg \, g^{-1}$) from an aqueous solution. In addition, SWCNTs and MWCNTs have better Ni(II) sorption properties following their oxidation with NaClO. These treatments improve polarity of the CNT surface, resulting in them becoming more hydrophilic and, therefore, able to sorb more charged metal ions from the aqueous solution [143, 144].

While the above-mentioned studies indicate that CNTs are potentially efficient adsorbents for a variety of pollutants in both drinking and environmental waters, their practical application may be hampered by their high cost [22]. However, CNTs could be applied at small amounts (thus at low cost) to the municipal sludge or to other solid wastes to absorb various organic contaminants so that these wastes could be safely land-applied to increase soil quality and minimize the waste-disposal expenses (see Sections 2.4 and 3).

The pristine CNTs are prone to aggregation and precipitation in the aqueous phase due to their extreme hydrophobicity [145, 146]. Dispersion of CNTs in the aqueous phase can be achieved either by modifying the surface structure and introducing hydrophilic (polar) functional groups [146, 147] or by improving the interactions on the nanotubes/water interface through addition of surfactants [148], polymers [149], and natural organic matter [145, 147, 150]. The former method directly enhances the hydrophility of the CNTS, while the latter options not only create a thermodynamically suitable surface in water but also provide steric or electrostatic repulsion among dispersed CNTs, thus preventing aggregation [145]. Natural organic matter may play important roles in fate and transport of nanotubes in the environment because of its ubiquitous presence. Hyung et al. [145] reported that the water samples taken from the Suwannee River, USA, showed a similar MWCNT stabilizing capacity as compared to fabricated solutions containing the model natural organic matter (SR-NOM). For the same initial MWCNT concentrations, the concentrations of suspended MWCNTs in SR-NOM solutions and the Suwannee River water samples were even considerably higher than that in a solution of 1% sodium dodecyl sulfate, a commonly used surfactant to stabilize CNTs in the aqueous phase.

Through studying the transport of carboxyl-functionalized SWCNTs in quartz-sand packed columns, Jaisi and Elimelech [146] and Jaisi et al. [147] concluded that the behaviors of the nanotubes were generally comparable to those traditionally observed with colloidal particles and bacterial cells. For example, an increase of the solution ionic strength resulted in increased SWCNT deposition in the column and divalent cations (e.g., Ca^{2+}) reduce the SWCNT stability much more effectively than monovalent cations (e.g., Na^+) at the same ionic strength. However, at very low ionic strengths even in DI water, SWCNT disposition in the sand media changed slightly, implying that the simply physical constrains (straining) also played roles in nanotube transport besides the complicated physicochemical interactions between particle and the medium surfaces. As concluded by Jaisi and Elimelech [146], straining may play more important roles on nanotube transport in the soil media. They compared the transport of linear nanotubes and spherical fullerene nanoparticles in columns packed with the same soils. It was found that the fullerene deposition rates were much lower than those of SWCNTs at the same ionic strength. Furthermore, fullerene nanoparticles were more sensitive to changes in ionic strength compared to SWCNTs. The authors proposed that linear shape and structure, particularly the very large aspect ratio and its highly bundled (aggregated) state in aqueous solutions, were mainly responsible for nanotube retentions in the soil columns. Moreover, the pore size distribution and pore geometry as well as heterogeneity in soil particle size, porosity, and permeability also contribute to straining in flow through the soil media by nanotubes. Thus, SWCNT transport in soils would be limited [147]. Similarly limited mobility was also reported on MWCNTs [151]. On the other hand, natural soil environments are more heterogeneous and normally contain open soil structures (e.g., cracks, fissures, worm trails, and other open features) that can promote preferential flow of SWNTs in soil. Additionally, soil pore water is normally rich in dissolved organic molecules (e.g., humic and fulvic acids) that can enhance the colloidal stability of nanomaterials [147].

Since there is limited number of publications of systemically studying the nanoparticles transport in the soil media, the discussions above showed significant implications on mobility of all types of nanoparticles in the soil environment. On one hand, nanoparticles may show decreased transport and higher retention rate in soil media than what were observed using sand-packed column tests in the laboratory due to the more complicated pore structures and pore distributions in soils. On the other hand, presence of the preferential flow and natural organic matter in soil media would enhance the nanoparticles mobility through the soil columns and increase the risks of groundwater contamination.

Studies have shown that CNTs are biologically active as demonstrated by a pulmonary response via induction of pulmonary granulomas [152, 153] at a greater instance than quartz (1–3 μm crystalline silica), which is a recognized chronic occupational health hazard (via inhalation routes). Both SWCNTs and MWCNTs were also attributed to cause loss of phagocytic ability and ultrastructure damage to alveola macrophages [154]. Furthermore, CNTs have induced observable toxic responses in other cell cultures [155, 156].

Using Nanoenhanced Materials as Solid Waste Stabilizers/Conditioners

Most solid wastes often contain environmental detrimental impurities, pathogens, and sometimes nauseous odors. Thus, beneficial reuses of such materials as resources in soil reclamation are limited by the concerns over secondary environmental contaminations. Through amendments, nanoenhanced materials might be able to increase the environmental safety and public acceptance of the waste application in reclaiming the mine soils or agricultural lands. For example, Li et al. [157] indicated that a small amount of nZVI (0.1% by weight) effectively removed the organic sulfur compounds (responsible for nuisance odors), heavy metals, and organic contaminants in the biosolids, suggesting nZVI could reduce the detrimental effect of biosolids and enhance beneficial uses of these organic and nutritious solids. Turan [158] observed that cocompost of poultry litter with natural zeolites at a ratio of 5% and 10% ($g\,g^{-1}$) resulted in 66% and 89% reduction of the end product salinity, respectively. Zeolites can absorb the toxic metals (100% of Cd, 28–45% of Cu, 10–15% of Cr, 50–55% of Ni and Pb, and 40–46% of Zn) in the biosolids at rates of 25%–30% ($g\,g^{-1}$) and reduce the leaching of these metals [159]. Nissen et al. [160] reported that addition of 0.5% and 1.0% zeolite over a 90-day period significantly reduced labile Zn in an experimental horticultural compost derived from sewage sludge. Subsequent plant growth trials measuring transfer of Zn and Cu to ryegrass in successive harvests demonstrated that 1.0% zeolite caused significant reduction in total metal transfer from soil to plant over a 116 d growth period. The use of zeolites is a cost-effective amendment for compost to significantly reduce potential for soil metal mobility and soil to plant transfer [160]. Villaseñor et al. [161] added three commercial natural zeolites to a pilot-scale rotary drum composting reactor, where the domestic sewage sludge and barley straws were cocomposted. They observed that all three types of zeolites removed 100% of Ni, Cr, Pb, and significant amounts (more than 60%) of Cu, Zn, and Hg originated from the sludge [161]. It is also reported that the clinoptilolites reduced 50% of the NH_3 emission from the compost [161], avoiding N loss and unpleasant odor from the compost. Villaseñoret al. [161] claimed that addition of 10% zeolites produced composts compliant with Spanish regulations regarding heavy metal contamination. According to them, the zeolite-amended compost could either be applied directly to soil, or the metal-polluted zeolites could be separated from the compost prior to application to ensure the environmental safety. Using zeolites as heavy metal absorbents in compost is also verified by other researchers [162–164]. Gadepalle et al. [165] applied compost containing 5% zeolite to an As-contaminated soil and observed that zeolites addition can effectively reduce the As uptake by rye grass and that less than 0.01% of the total As content in the soil may be absorbed by the plants.

Literature above showed that amending the solid wastes with relatively small amounts of nanomaterials could effectively reduce or eliminate the risk of secondary contamination associated with land applications of these wastes. This practice could expand the industrial or municipal waste lists which are safe for land application, thus saving the cost of waste disposal and ameliorating the adverse environmental impacts. In addition, agricultural soils and drastically disturbed lands (e.g., mine soils) could benefit from these most cost-effective waste materials (soil amendments). Moreover, application of the nanomaterials to stabilize or condition the conventional soil amendment materials (e.g., composts, biosolids, coal combustion by-products) could be a potential aspect of utilization of nanotechnology in the agriculture at low cost. Zeolites, nFeOs, phosphate-based nanoparticles, and sulfide-based nanoparticles are efficient in immobilizing inorganic contaminants in the solids, while C nanotubes have a high absorption capacity for organic pollutants and nZVI can destroy the OWCs present in the wastes by reduction reactions. Finally, incubation of the nanomaterials with solid wastes could in turn stabilize the former and reduce the risks of nanomaterials spill and contaminations resulting from direct application of the nanoparticles to the environment.

Using Nanoenhanced Materials to Control Soil Erosion

Soil erosion caused by rainfall or wind in a closed mining site, especially before vegetation is established, can result in loss of good soil, exposure of the buried sulfide minerals, and transportation of the sediments and contaminants to the nearby surface water bodies. Thus, soil erosion control is a high priority in a mine soil reclamation plan. Nanoenhanced materials have been used to combat the soil erosion problems. Andry et al. [166] applied 10% of a Ca-type zeolite material to an acidic soil and tested surface runoff and soil loss under simulated rainfall. They observed that the surface runoff and soil loss can be substantially reduced by zeolite application because of an increase in wet aggregate stability and the large particle size of the sediment due to the amendments. The authors [166] claimed that zeolites can be more effective than lime in control soil erosion. Yamamoto et al. [167] also observed the decrease of runoff rate and soil loss in sodic soils mixed with Ca type of artificial zeolite at rates of 5–25%. They assumed that the exchange of Ca^{2+} on zeolites with Na^+ in the sodic soil reduced the clay dispersion, resulting in increased soil hydraulic conductivity and soil aggregation. Zheng [168] applied polyacrylamide (PAM, a polyelectrolyte used for soil erosion control) and magnetite nanoparticles to an As-spiked soil subject to the simulated rainfall and concluded that PAM could effectively reduce soil erosion while the nanoparticles could reduce As leaching. Wang et al. [169] examined the effects of alumina nanoparticles (Al_2O_3, 140–330 nm) and a cationic polyelectrolyte in conditioning a wastewater treatment sludge and reported that this amendment can result in larger flocs and better dewatering effects than the single conditioning by polyelectrolyte only. The beneficial effects are more pronounced when finer nanoparticles (140 nm) were used. Wang and colleagues [169] proposed that the nanoparticles can enhance the stretch of the chain-like structures of the polyelectrolyte, resulting in more effective bridging effects and better flocculation. As a matter of fact, the PE (polyelectrolyte)—NP (nanoparticles) flocculation systems have been widely used in effectively removing solid particles from the solution [170, 171]. The flocculation in such a system is induced by the sequential addition of a positively charged polyelectrolyte followed by negatively charged nanoparticles, such as bentonite and colloidal silica. The systems produce a better flocculation and drainage (dewatering) than conventional polymer-only flocculation systems [170]. These results suggest that dual application of polyelectrolyte and nanoparticles could enhance flocculation and increase soil particle size and particle stability and thus effectively control soil erosions caused by wind or rain.

SUMMARY: TOWARD A PRACTICAL STRATEGY IN APPLYING NANO-TECHNOLOGY IN MINE SOILRECLAMATION AND AGRICULTURE

Mine soil reclamation and rehabilitation can benefit from local environmental protection, provide additional land for forest or agricultural uses, and offset the atmospheric CO_2 increase through C sequestration. Yet, the drastically disturbed land surfaces usually have poor soil quality where the desired plant species such as trees and crops are difficult to establish. Although some industrial, agricultural, and municipal wastes could be applied to the mining sites as soil amendments to improve soil quality, environmentally sensitive impurities contained in these wastes hinder the wide application of such materials. A synthesis of literature presented in this paper supports the hypothesis that the nanoenhanced materials, which have been successfully used in industry and other areas as new emerging materials with unique properties, could also be used as amendments to improve the quality of mine and agricultural soils at high efficiency. Although the data in support of this hypothesis are limited, the literature indicates that zeolites, nanoiron oxide particles, nanozero valent iron particles, nano-phosphate-based particles, nanosulfide-based particles, carbon nanotubes can improve soil physical and chemical properties, enhance soil fertility, stabilize soil contaminants, or reduce soil erosion. Mobility of nanoparticles in the soil columns is generally limited and spill of the nanomaterials should not be an issue although preferential flow and the dissolved organic matter might enhance mobility and transport of the particles. Iron-based nanoparticles, derived from environmental friendly Fe compounds, generally have low ecotoxicological effects. Thus, these nanoparticles are promising candidates for use in mine soil reclamation as well as for improving quality of agricultural soils.

One of the most important factors impeding a wide application of nanotechnology in agriculture is the cost of the nano-materials. However, natural nano-materials such as zeolites are usually not very costly. The average price for clinoptilolite granules is about $145 per

Mg (tonne), and some of the modified clinoptilolite and activated chabazite products are sold for as much as $8 per kg [172]. For some other nanoparticles, which could be made in situ, the price depends on the chemicals used to synthesize the particles. The cost of nZVI is about $50–100 kg^{-1} [173], and nanoiron oxides, nanoiron phosphates, and nanoiron sulfides should be in the similar price range since the fabrication methods are similar. In comparison, CNTs are relatively expensive. The retail price of SWCNTs was $50 per g in the year of 2000 [174]. However, the prices of nanoparticles are always higher compared with the conventional soil amendment materials such as fly ash, manures, composts, or biosolids, which are often free to use.

Therefore, a practical strategy is proposed to make a feasible use of the nanotechnology in soil reclamation. The strategy is to (a) mix the suitable nanoparticles with the conventional amendment materials at small quantities, (b) stabilize the solid wastes to a certain degree and eliminate the risks of secondary contaminations by absorbing/immobilizing the heavy metals and organic toxins, and (c) apply the nanomaterial-amended wastes to mine soils or agricultural soils for better crop/vegetation establishments. Because of high effectiveness of the nanoparticles, 1–10% nanoparticles by weight are usually adequate to minimize the contamination problems associated with in the solid wastes. For instance, assuming a rate of 10 Mg ha^{-1} of biosolids is applied to a mining site for reclamation, according to [157], only 0.1% of zNVI is needed to stabilize the bio-solid [157]. In other words, only 10 kg ha^{-1} nZVI is required, which is a cost-effective rate of application of nanoparticles as soil amendment. Therefore, application of nanoparticle-stabilized conventional soil amendment materials is a practical approach to use nanotechnology for mine soil reclamation and for agriculture. Meanwhile, incubation of the highly reactive nanoparticles with solid wastes could reduce the risks of direct release of the nanoparticles to the environment and toxicity of nanoparticles to the plants.

However, there is few research or practical examples of using nanotechnology for soil reclamation. Therefore, assessing the effectiveness of nanoparticles in mine soil reclamation and vegetation establishments is a researchable priority. A high priority must also be given to studying the feasibility of using nanoparticle-stabilized solid wastes and of developing nanofertilizers.

ACKNOWLEDGMENTS

This project was funded by the Office of Ohio Development.

REFERENCES

1. P. L. Younger, "Environmental impacts of coal mining and associated wastes: a geochemical perspective," Geological Society Special Publication, no. 236, pp. 169–209, 2004.
2. F. G. Bell and L. J. Donnelly, Mining and Its Impact on the Environment, Taylor & Francis, New York, NY, USA, 2006.
3. S. F. Greb, C. F. Eble, D. C. Peters, and A. R. Papp, Coal and the Environment, American Geological Institute, Alexandria, Va, USA, 2006.
4. W. L. Daniels, B. Stewart, and C. E. Zipper, Reclamation of Coal Refuse Disposal Areas VCE publication 460-131, www.http://ww.pubs.ext.vt.edu/460-131.html, 2010.
5. D. A. N. Ussiri and R. Lal, "Carbon sequestration in reclaimed minesoils," Critical Reviews in Plant Sciences, vol. 24, no. 3, pp. 151–165, 2005. · ·
6. M. Sperow, "Carbon sequestration potential in reclaimed mine sites in seven east-central states," Journal of Environmental Quality, vol. 35, no. 4, pp. 1428–1438, 2006. · ·
7. M. Pietrzykowski and W. Krzaklewski, "Potential for carbon sequestration in reclaimed mine soil on reforested surface mining areas in Poland," Natural Science, vol. 2, pp. 1015–1021, 2010.
8. R. K. Shrestha, R. Lal, and P. A. Jacinthe, "Enhancing carbon and nitrogen sequestration in reclaimed soils through organic amendments and chiseling," Soil Science Society of America Journal, vol. 73, no. 3, pp. 1004–1011, 2009. · ·
9. K. C. Haering, W. L. Daniels, and S. E. Feagley, "Reclaiming mined lands with biosolids, manures and papermill sludges," in Reclamation of Drastically Disturbed Lands, R. I. Barnhisel, R. G. Darmody, and W. L. Daniels, Eds., pp. 615–644, American Society of Agronomy, Soil Science Society of America, Soil Science Society of America, Madison, Wis, USA, 2000.

10.   F. J. Larney, O. O. Akinremi, R. L. Lemke, V. E. Klaassen, and H. H. Janzen, "Soil responses to topsoil replacement depth and organic amendments in wellsite reclamation," Canadian Journal of Soil Science, vol. 85, no. 2, pp. 307–317, 2005.

11.   L. S. Forsberg and S. Ledin, "Effects of sewage sludge on pH and plant availability of metals in oxidising sulphide mine tailings," Science of the Total Environment, vol. 358, no. 1–3, pp. 21–35, 2006. · ·

12.   E. S. Bendfeldt, J. A. Burger, and W. Lee Daniels, "Quality of amended mine soils after sixteen years," Soil Science Society of America Journal, vol. 65, no. 6, pp. 1736–1744, 2001.

13.   D. W. Ming and E. R. Allen, "Use of natural zeolites in agronomy, horticulture and environmental soil remediation," in Natural Zeolites: Occurrence, Properties, Applications, D. W. Ming and D. B. Bish, Eds., pp. 619–654, Mineralogical Society of America, Geochemical Society, Saint Louis, Mo, USA; Italian National Academy, Accademia Nationale dei Lincei (ANL), Barcelona, Italy, 2001.

14.   W. A. Dick, R. C. Stehouwer, J. M. Bigham, et al., "Beneficial uses of flue gas desulfurization by-products: examples and case studies of land application," in Land Application of Agricultural, Industrial, and Municipal By-Products, J. F. Power and W. A. Dick, Eds., pp. 505–536, Soil Science Society of America, Madison, Wis, USA, 2000.

15.   D. K. Bhumbla, R. N. Singh, and R. F. Keefer, "Coal combustion by-product utilization for land reclamation," in Reclamation of Drastically Disturbed Lands, R. I. Barnhisel, R. G. Darmody, and W. L. Daniels, Eds., pp. 489–512, American Society of Agronomy, Soil Science Society of America, Soil Science Society of America, Madison, Wis, USA, 2000.

16.   W. L. Daniels, B. Stewart, K. C. Haering, and C. E. Zipper, "The potential for beneficial reuse of coal fly ash in southwest Virginia mining environments," Publication 460-134, Virginia Cooperative Extension (VCE), Stanardsville, Va, USA, 2002.

17.   P. A. Jacinthe and R. Lal, "Carbon storage and minesoil properties in relation to topsoil application techniques," Soil Science Society of America Journal, vol. 71, no. 6, pp. 1788–1795, 2007. · ·

18. R. B. Clark, K. D. Ritchey, and V. C. Baligar, "Benefits and constraints for use of FGD products on agricultural land," Fuel, vol. 80, no. 6, pp. 821–828, 2001. · ·

19. G. A. O›Connor, H. A. Elliott, N. T. Basta et al., "Sustainable land application: an overview,"Journal of Environmental Quality, vol. 34, no. 1, pp. 7–17, 2005.

20. C. A. Kinney, E. T. Furlong, S. D. Zaugg et al., "Survey of organic wastewater contaminants in biosolids destined for land application," Environmental Science and Technology, vol. 40, no. 23, pp. 7207–7215, 2006. · ·

21. J. Kim, "Preface," in Advances in Nanotechnology And the Environment, J. Kim, Ed., Pan Stanford Publishing, Singapore, Singapore, 2012.

22. J. Theron, J. A. Walker, and T. E. Cloete, "Nanotechnology and water treatment: applications and emerging opportunities," Critical Reviews in Microbiology, vol. 34, no. 1, pp. 43–69, 2008. · ·

23. M. S. Mauter and M. Elimelech, "Environmental applications of carbon-based nanomaterials," Environmental Science and Technology, vol. 42, no. 16, pp. 5843–5859, 2008. · ·

24. T. Masciangioli and W. X. Zhang, "Environmental technologies at the nanoscale,"Environmental Science and Technology, vol. 37, no. 5, 2003.

25. R. Lal, "Promise and limitations of soils to minimize climate change," Journal of Soil and Water Conservation, vol. 63, no. 4, 2008.

26. R. T. Pabalan and F. P. Bertetti, "Cation-exchange properties of natural zeolites," in Natural Zeolites: Occurrence, Properties, Applications, D. L. Bish and D. W. Ming, Eds., vol. 45, pp. 453–518, Mineralogical Society of America Reviews in Mineralogy and Geochemistry, Washington, DC, USA, 2001.

27. F. A. Mumpton, "Using zeolites in agriculture," in Innovative Biological Technologies for Lesser Developed Countries, Congress of the United States, Office of Technology Assessment, Washington, DC, USA, 1985.

28. J. L. Boettinger and D. W. Ming, "Zeolites," in Soil Mineralogy with Environmental Applications, J. B. Dixon and D. G. Schulze,

Eds., SSSA Book Series 7, pp. 585–610, Soil Science Society of America, Madison, Wis, USA, 2002.

29. L. J. M. Githinji, J. H. Dane, and R. H. Walker, "Physical and hydraulic properties of inorganic amendments and modeling their effects on water movement in sand-based root zones," Irrigation Science, vol. 29, no. 1, pp. 65–77, 2011. · ·

30. G. R. Wehtje, J. N. Shaw, R. H. Walker, and W. Williams, "Bermudagrass growth in soil supplemented with inorganic amendments," HortScience, vol. 38, no. 4, pp. 613–617, 2003.

31. A. M. Petrovic, "The potential of natural zeolite as a soil amendment," Golf Course Manage, vol. 58, no. 11, pp. 92–93, 1990.

32. Z. T. Huang and A. M. Petrovic, "Physical properties of sand as affected by clinoptilolite zeolite particle size and quantity," Journal of Turfgrass Management, vol. 1, no. 1, pp. 1–15, 1995.

33. Z. T. Huang and A. M. Petrovic, "Clinoptilolite zeolite effect on evapotranspiration rate and shoot growth rate of creeping bentgrass on sand base greens," Journal of Turfgrass Management, vol. 1, no. 4, pp. 1–9, 1996.

34. Z. Lopez, A. S. Bawazir, B. Tanzy, and E. Adkins, "Using St. Cloud clinoptilolite zeolite as a wicking material to sustain riparian vegetation," in Proceedings of the 2008 Joint Meeting of The Geological Society of America, Soil Science Society of America, American Society of Agronomy, Crop Science Society of America, Gulf Coast Association of Geological Societies with the Gulf Coast Section of SEPM. Paper No. 54-6, 2008.

35. Z. T. Huang and A. M. Petrovic, "Clinoptilolite zeolite influence on nitrate leaching and nitrogen use efficiency in simulated sand based golf greens," Journal of Environmental Quality, vol. 23, no. 6, pp. 1190–1194, 1994.

36. L. E. Katz, D. N. Humphrey, P. T. Jankauskas, and F. A. Demascio, "Engineered soils for low-level radioactive waste disposal facilities: effects of additives on the adsorptive behavior and hydraulic conductivity of natural soils," Hazardous Waste and Hazardous Materials, vol. 13, no. 2, pp. 283–306, 1996.

37. H. Khan, A. Z. Khan, R. Khan, N. Matsue, and T. Henmi, "Influence of zeolite application on germination and seed quality

of soybean grown on allophanic soil," Research Journal of Seed Science, vol. 2, no. 1, pp. 1–8, 2009. · ·

38. K. Torii, "Utilization of natural zeolites in Japan," in Natural Zeolites: Occurrence, Properties, Use, L. B. Sand and F. A. Mumpton, Eds., pp. 441–450, Pergamon Press, Elmsford, NY, USA, 1978.

39. G. A. Mazur, G. K. Medvid, and T. I. Grigora, "Use of natural zeolites for increasing the fertility of light textured soils," Eurasian Soil Science, vol. 10, pp. 70–77, 1984.

40. R. Liu and R. Lal, "A laboratory study on improvement of mine soil quality for re-vegetation through various amendments," in Proceedings of the ASA-CSSA-SSSA International Annual Meetings, Cincinnati, Ohio, USA, October 2012.

41. J. A. Burger and C. E. Zipper, "How to restore forests on surface-mined land," Publication460-123, Virginia Cooperative Extension (VCE), Stanardsville, Va, USA, 2011.

42. T. S. Perrin, D. T. Drost, J. L. Boettinger, and J. M. Norton, "Ammonium-loaded clinoptilolite: a slow-release nitrogen fertilizer for sweet corn," Journal of Plant Nutrition, vol. 21, no. 3, pp. 515–530, 1998.

43. M. D. Lewis, I. F. D. Moore, and K. L. Goldsberry, "Ammonium-exchanged clinoptilolite and granulated clinoptilolite with urea as nitrogen fertilizers," in Zeo-Agriculture: Use of Natural Zeolites in Agriculture and Aquaculture, W. G. Pond and F. A. Mumpton, Eds., pp. 105–111, Westview Press, Boulder, Colo, USA, 1984.

44. K. A. Barbarick and H. J. Pirela, "Agronomic and horticultural uses of zeolites: a review," inZeo-Agriculture: Use of Natural Zeolites in Agriculture and Aquaculture, W. G. Pond and F. A. Mumpton, Eds., pp. 93–103, Westview Press, Boulder, Colo, USA, 1984.

45. K. A. Williams and P. V. Nelson, "Using precharged zeolite as a source of potassium and phosphate in a soilless container medium during potted chrysanthemum production,"Journal of the American Society for Horticultural Science, vol. 122, no. 5, pp. 703–708, 1997.

46. J. L. Carlino, K. A. Williams, and E. R. Allen, "Evaluation of zeolite-based soilless root media for potted chrysanthemum production," HortTechnology, vol. 8, no. 3, pp. 373–378, 1998.

47. T. M. Lai and D. D. Eberl, "Controlled and renewable release of phosphorous in soils from mixtures of phosphate rock and NH4-exchanged clinoptilolite," Zeolites, vol. 6, no. 2, pp. 129–132, 1986.

48. D. D. Eberl, K. A. Barbarick, and T. M. Lai, "Influence of NH4-exchanged clinoptilolite on nutrient concentrations in sorghum-sudangrass," in Natural Zeolites ‹93: Occurrence, Properties, Use, D. W. Ming and F. A. Mumpton, Eds., pp. 491–504, Int›l Comm Natural Zeolites, Brockport, NY, USA, 1995.

49. E. R. Allen, L. R. Hossner, D. W. Ming, and D. L. Henninger, "Solubility and cation exchange in phosphate rock and saturated clinoptilolite mixtures," Soil Science Society of America Journal, vol. 57, no. 5, pp. 1368–1374, 1993.

50. Z. L. He, V. C. Baligar, D. C. Martens, K. D. Ritchey, and M. Elrashidi, "Effect of byproduct, nitrogen fertilizer, and zeolite on phosphate rock dissolution and extractable phosphorus in acid soil," Plant and Soil, vol. 208, no. 2, pp. 199–207, 1999. · ·

51. M. C. Derosa, C. Monreal, M. Schnitzer, R. Walsh, and Y. Sultan, "Nanotechnology in fertilizers," Nature Nanotechnology, vol. 5, no. 2, p. 91, 2010. · ·

52. R. Liu and D. Zhao, "Reducing leachability and bioaccessibility of lead in soils using a new class of stabilized iron phosphate nanoparticles," Water Research, vol. 41, no. 12, pp. 2491–2502, 2007. · ·

53. R. Liu, "In-situ lead remediation in a shoot-range soil using stabilized apatite nanoparticles," in Proceedings of the 85th ACS Colloid and Surface Science Symposium, McGill University, Montreal, Canada, June 2011.

54. R. Edwards, I. Rebedea, N. W. Lepp, and A. J. Lovell, "An investigation into the mechanism by which synthetic zeolites reduce labile metal concentrations in soils," Environmental Geochemistry and Health, vol. 21, no. 2, pp. 157–173, 1999. · ·

55. C. F. Lin, S. S. Lo, H. Y. Lin, and Y. Lee, "Stabilization of cadmium contaminated soils using synthesized zeolite," Journal of Hazardous Materials, vol. 60, no. 3, pp. 217–226, 1998. · ·

56. A. Shanableh and A. Kharabsheh, "Stabilization of Cd, Ni and Pb in soil using natural zeolite," Journal of Hazardous Materials, vol. 45, no. 2-3, pp. 207–217, 1996. · ·

57.   A. Moirou, A. Xenidis, and I. Paspaliaris, "Stabilization Pb, Zn, and Cd-contaminated soil by means of natural zeolite," Soil and Sediment Contamination, vol. 10, no. 3, pp. 251–267, 2001.

58.   C. Haidouti, "Inactivation of mercury in contaminated soils using natural zeolites," Science of the Total Environment, vol. 208, no. 1-2, pp. 105–109, 1997. · ·

59.   A. Chlopecka and D. C. Adriano, "Mimicked in-situ stabilization of metals in a cropped soil: bioavailability and chemical form of zinc," Environmental Science and Technology, vol. 30, no. 11, pp. 3294–3303, 1996. · ·

60.   A. S. Knox, D. I. Kaplan, D. C. Adriano, T. G. Hinton, and M. D. Wilson, "Apatite and phillipsite as sequestering agents for metals and radionuclides," Journal of Environmental Quality, vol. 32, no. 2, pp. 515–525, 2003.

61.   M. R. Mahmoodabadi, "Experimental study on the effects of natural zeolite on lead toxicity, growth, nodulation, and chemical composition of soybean," Communications in Soil Science and Plant Analysis, vol. 41, no. 16, pp. 1896–1902, 2010. · ·

62.   W. Geebelen, J. Vangronsveld, D. C. Adriano, R. Carleer, and H. Clijsters, "Amendment-induced immobilization of lead in a lead-spiked soil: evidence from phytotoxicity studies,"Water, Air, and Soil Pollution, vol. 140, no. 1–4, pp. 261–277, 2002. · ·

63.   E. Coppola, G. Battaglia, M. Bucci et al., "Remediation of Cd- and Pb-polluted soil by treatment with organo-zeolite conditioner," Clays and Clay Minerals, vol. 51, no. 6, pp. 609–615, 2003. · ·

64.   K. Stead, Environmental implications of using the natural zeolite clinoptilolite for the remediation of sludge amended soils [Ph.D. thesis], University of Surrey, Surrey, UK, 2002.

65.   J. M. Bigham, R. W. Fitzpatrick, and D. G. Schulze, "Iron oxides," in Soil Mineralogy with Environmental Applications, J. B. Dixon and D. G. Schulze, Eds., pp. 323–366, Soil Science Society of America, Madison, Wis, USA, 2002.

66.   M. Hua, S. Zhang, B. Pan, W. Zhang, L. Lv, and Q. Zhang, "Heavy metal removal from water/wastewater by nanosized metal oxides: a review," Journal of Hazardous Materials, vol. 211-212, pp. 317–331, 2012.

67. A. Xenidis, C. Stouraiti, and N. Papassiopi, "Stabilization of Pb and As in soils by applying combined treatment with phosphates and ferrous iron," Journal of Hazardous Materials, vol. 177, no. 1–3, pp. 929–937, 2010. · ·

68. J. Kumpiene, A. Lagerkvist, and C. Maurice, "Stabilization of As, Cr, Cu, Pb and Zn in soil using amendments-a review," Waste Management, vol. 28, no. 1, pp. 215–225, 2008. · ·

69. USEPA, the Use of Soil Amendments for Remediation, Revitalization and Reuse. Solid Waste and Emergency Response (5203P) EPA 542-R-07-013, http://clu-in.org/download/remed/epa-542-r-07-013.pdf, 2007.

70. H. J. Shipley, K. E. Engates, and A. M. Guettner, "Study of iron oxide nanoparticles in soil for remediation of arsenic," Journal of Nanoparticle Research, vol. 13, no. 6, pp. 2387–2397, 2011.

71. D. O'Carroll, B. Sleep, M. Krol, H. Boparai, and C. Kocur, "Nanoscale zero valent iron and bimetallic particles for contaminated site remediation," Advances in Water Resources. In press. ·

72. J. D. Hu, Y. Zevi, X. M. Kou, J. Xiao, X. J. Wang, and Y. Jin, "Effect of dissolved organic matter on the stability of magnetite nanoparticles under different pH and ionic strength conditions," Science of the Total Environment, vol. 408, no. 16, pp. 3477–3489, 2010. · ·

73. Y. T. He, J. Wan, and T. Tokunaga, "Kinetic stability of hematite nanoparticles: the effect of particle sizes," Journal of Nanoparticle Research, vol. 10, no. 2, pp. 321–332, 2008. · ·

74. M. Baalousha, "Aggregation and disaggregation of iron oxide nanoparticles: influence of particle concentration, pH and natural organic matter," Science of the Total Environment, vol. 407, no. 6, pp. 2093–2101, 2009. · ·

75. Y. Hong, R. J. Honda, N. V. Myung, and S. L. Walker, "Transport of iron-based nanoparticles: role of magnetic properties," Environmental Science and Technology, vol. 43, no. 23, pp. 8834–8839, 2009. ·

76. M. Baalousha, A. Manciulea, S. Cumberland, K. Kendall, and J. R. Lead, "Aggregation and surface properties of iron oxide nanoparticles: influence of pH and natural organic

matter,"Environmental Toxicology and Chemistry, vol. 27, no. 9, pp. 1875–1882, 2008. · ·

77. H. L. Karlsson, J. Gustafsson, P. Cronholm, and L. Möller, "Size-dependent toxicity of metal oxide particles-A comparison between nano- and micrometer size," Toxicology Letters, vol. 188, no. 2, pp. 112–118, 2009. · ·

78. M. Auffan, L. Decome, J. Rose et al., "In vitro interactions between DMSA-coated maghemite nanoparticles and human fibroblasts: a physicochemical and cyto-genotoxical study,"Environmental Science and Technology, vol. 40, no. 14, pp. 4367–4373, 2006. · ·

79. L. K. Limbach, P. Wick, P. Manser, R. N. Grass, A. Bruinink, and W. J. Stark, "Exposure of engineered nanoparticles to human lung epithelial cells: influence of chemical composition and catalytic activity on oxidative stress," Environmental Science and Technology, vol. 41, no. 11, pp. 4158–4163, 2007. · ·

80. N. Sadeghiani, L. S. Barbosa, L. P. Silva, R. B. Azevedo, P. C. Morais, and Z. G. M. Lacava, "Genotoxicity and inflammatory investigation in mice treated with magnetite nanoparticles surface coated with polyaspartic acid," Journal of Magnetism and Magnetic Materials, vol. 289, pp. 466–468, 2005. · ·

81. W. X. Zhang, "Nanoscale iron particles for environmental remediation: an overview," Journal of Nanoparticle Research, vol. 5, no. 3-4, pp. 323–332, 2003. ·

82. B. Cao, B. Ahmed, and H. Beyenal, "Immobilization of uranium in groundwater using biofilms," in Emerging Environmental Technologies, V. Shah, Ed., vol. 2, pp. 1–37, Springer, New York, NY, USA, 2010.

83. A. Abdelouas, "Uranium mill tailings: geochemistry, mineralogy, and environmental impact," Elements, vol. 2, no. 6, pp. 335–341, 2006. · ·

84. S. Yan, B. Hua, Z. Bao, J. Yang, C. Liu, and B. Deng, "Uranium(VI) removal by nanoscale zerovalent iron in anoxic batch systems," Environmental Science and Technology, vol. 44, no. 20, pp. 7783–7789, 2010. · ·

85. J. N. Fiedor, W. D. Bostick, R. J. Jarabek, and J. Farrell, "Understanding the mechanism of uranium removal from

groundwater by zero- valent iron using X-ray photoelectron spectroscopy," Environmental Science and Technology, vol. 32, no. 10, pp. 1466–1473, 1998. · ·

86. R. A. Crane, M. Dickinson, I. C. Popescu, and T. B. Scott, "Magnetite and zero-valent iron nanoparticles for the remediation of uranium contaminated environmental water," Water Research, vol. 45, no. 9, pp. 2931–2942, 2011. · ·

87. M. Dickinson and T. B. Scott, "The application of zero-valent iron nanoparticles for the remediation of a uranium-contaminated waste effluent," Journal of Hazardous Materials, vol. 178, no. 1–3, pp. 171–179, 2010. · ·

88. O. Riba, T. B. Scott, K. Vala Ragnarsdottir, and G. C. Allen, "Reaction mechanism of uranyl in the presence of zero-valent iron nanoparticles," Geochimica et Cosmochimica Acta, vol. 72, no. 16, pp. 4047–4057, 2008. · ·

89. D. V. Franco, L. M. Da Silva, and W. F. Jardim, "Reduction of hexavalent chromium in soil and ground water using zero-valent iron under batch and semi-batch conditions," Water, Air, and Soil Pollution, vol. 197, no. 1–4, pp. 49–60, 2009. · ·

90. Y. Xu and D. Zhao, "Reductive immobilization of chromate in water and soil using stabilized iron nanoparticles," Water Research, vol. 41, no. 10, pp. 2101–2108, 2007. · ·

91. S. M. Ponder, J. G. Darab, and T. E. Mallouk, "Remediation of Cr(VI) and Pb(II) aqueous solutions using supported, nanoscale zero-valent iron," Environmental Science and Technology, vol. 34, no. 12, pp. 2564–2569, 2000. · ·

92. A. D. Lemly, "Environmental implications of excessive selenium: a review," Biomedical and Environmental Sciences, vol. 10, no. 4, pp. 415–435, 1997.

93. C. L. Mackowiak and M. C. Amacher, "Soil sulfur amendments suppress selenium uptake by alfalfa and western wheatgrass," Journal of Environmental Quality, vol. 37, no. 3, pp. 772–779, 2008. · ·

94. S. T. Witte and L. A. Will, "Investigation of selenium sources associated with chronic selenosis in horses of western Iowa," Journal of Veterinary Diagnostic Investigation, vol. 5, no. 1, pp. 128–131, 1993.

95. P. Thomas, J. Irvine, J. Lyster, and R. Beaulieu, "Radionuclides and trace metals in Canadian moose near uranium mines: Comparison of radiation doses and food chain transfer with cattle and caribou," Health Physics, vol. 88, no. 5, pp. 423–438, 2005.

96. J. T. Olegario, N. Yee, M. Miller, J. Sczepaniak, and B. Manning, "Reduction of Se(VI) to Se(-II) by zerovalent iron nanoparticle suspensions," Journal of Nanoparticle Research, vol. 12, no. 6, pp. 2057–2068, 2010. · ·

97. L. Alidokht, A. R. Khataee, A. Reyhanitabar, and S. Oustan, "Cr(VI) Immobilization process in a Cr-spiked soil by zerovalent iron nanoparticles: optimization using response surface methodology," Clean-Soil, Air, Water, vol. 39, no. 7, pp. 633–640, 2011. · ·

98. X. Q. Li and W. X. Zhang, "Sequestration of metal cations with zerovalent iron nanoparticles: a study with high resolution x-ray photoelectron spectroscopy (HR-XPS)," Journal of Physical Chemistry C, vol. 111, no. 19, pp. 6939–6946, 2007. · ·

99. X. Q. Li and W. X. Zhang, "Iron nanoparticles: the core-shell structure and unique properties for Ni(II) sequestration," Langmuir, vol. 22, no. 10, pp. 4638–4642, 2006. · ·

100. S. R. Kanel, J. M. Greneche, and H. Choi, "Arsenic(V) removal from groundwater using nano scale zero-valent iron as a colloidal reactive barrier material," Environmental Science and Technology, vol. 40, no. 6, pp. 2045–2050, 2006. · ·

101. S. R. Kanel, B. Manning, L. Charlet, and H. Choi, "Removal of arsenic(III) from groundwater by nanoscale zero-valent iron," Environmental Science and Technology, vol. 39, no. 5, pp. 1291–1298, 2005. · ·

102. T. Watanabe, Y. Murata, T. Nakamura, Y. Sakai, and M. Osaki, "Effect of zero-valent iron application on cadmium uptake in rice plants grown in cadmium-contaminated soils,"Journal of Plant Nutrition, vol. 32, no. 7, pp. 1164–1172, 2009. · ·

103. N. Saleh, H. Kim, T. Phenrat, et al., "Ionic strength and composition affect the mobility of surface-modified Fe0 nanoparticles in water-saturated sand columns," Environmental Science & Technology, vol. 42, no. 9, pp. 3349–3355, 2008. ·

104. P. G. Tratnyek and R. L. Johnson, "Nanotechnologies for environmental cleanup," Nano Today, vol. 1, no. 2, pp. 44–48, 2006. · ·

105. F. He and D. Zhao, "Preparation and characterization of a new class of starch-stabilized bimetallic nanoparticles for degradation of chlorinated hydrocarbons in water,"Environmental Science and Technology, vol. 39, no. 9, pp. 3314–3320, 2005. ·

106. F. He and D. Zhao, "Manipulating the size and dispersibility of zerovalent iron nanoparticles by use of carboxymethyl cellulose stabilizers," Environmental Science and Technology, vol. 41, no. 17, pp. 6216–6221, 2007. ·

107. T. Phenrat, N. Saleh, K. Sirk, H. J. Kim, R. D. Tilton, and G. V. Lowry, "Stabilization of aqueous nanoscale zerovalent iron dispersions by anionic polyelectrolytes: adsorbed anionic polyelectrolyte layer properties and their effect on aggregation and sedimentation," Journal of Nanoparticle Research, vol. 10, no. 5, pp. 795–814, 2008. · ·

108. N. Sakulchaicharoen, D. M. O›Carroll, and J. E. Herrera, "Enhanced stability and dechlorination activity of pre-synthesis stabilized nanoscale FePd particles," Journal of Contaminant Hydrology, vol. 118, no. 3-4, pp. 117–127, 2010. ·

109. B. C. Reinsch, B. Forsberg, R. L. Penn, C. S. Kim, and G. V. Lowry, "Chemical transformations during aging of zerovalent iron nanoparticles in the presence of common groundwater dissolved constituents," Environmental Science and Technology, vol. 44, no. 9, pp. 3455–3461, 2010. · ·

110. K. D. Grieger, A. Fjordbøge, N. B. Hartmann, E. Eriksson, P. L. Bjerg, and A. Baun, "Environmental benefits and risks of zero-valent iron nanoparticles (nZVI) for in situ remediation: risk mitigation or trade-off?" Journal of Contaminant Hydrology, vol. 118, no. 3-4, pp. 165–183, 2010. · ·

111. M. V. Ruby, A. Davis, and A. Nicholson, "In situ formation of lead phosphates in soils as a method to immobilize lead," Environmental Science Technology, vol. 28, no. 4, pp. 646–654, 1994.

112. Q. Y. Ma, T. J. Logan, and S. J. Traina, "Lead immobilization from aqueous solutions and contaminated soils using phosphate rocks," Environmental Science and Technology, vol. 29, no. 4, pp. 1118–1126, 1995.

113. S. Raicevic, T. Kaludjerovic-Radoicic, and A. I. Zouboulis, "In situ stabilization of toxic metals in polluted soils using phosphates: theoretical prediction and experimental verification," Journal of Hazardous Materials, vol. 117, no. 1, pp. 41–53, 2005. · ·

114. S. Raicevic, J. V. Wright, V. Veljkovic, and J. L. Conca, "Theoretical stability assessment of uranyl phosphates and apatites: selection of amendments for in situ remediation of uranium," Science of the Total Environment, vol. 355, no. 1–3, pp. 13–24, 2006. · ·

115. N. T. Basta and S. L. McGowen, "Evaluation of chemical immobilization treatments for reducing heavy metal transport in a smelter-contaminated soil," Environmental Pollution, vol. 127, no. 1, pp. 73–82, 2004. · ·

116. T. T. Eighmy, B. S. Crannell, L. G. Butler et al., "Heavy metal stabilization in municipal solid waste combustion dry scrubber residue using soluble phosphate," Environmental Science and Technology, vol. 31, no. 11, pp. 3330–3338, 1997. · ·

117. R. Stanforth and J. Qiu, "Effect of phosphate treatment on the solubility of lead in contaminated soil," Environmental Geology, vol. 41, no. 1-2, pp. 1–10, 2001. · ·

118. M. Peld, K. Tõnsuaadu, and V. Bender, "Sorption and desorption of Cd^{2+} and Zn^{2+} ions in apatite-aqueous systems," Environmental Science and Technology, vol. 38, no. 21, pp. 5626–5631, 2004. ·

119. A. S. Knox, D. I. Kaplan, and M. H. Paller, "Phosphate sources and their suitability for remediation of contaminated soils," Science of the Total Environment, vol. 357, no. 1–3, pp. 271–279, 2006. · ·

120. USEPA, US Environmental Protection Agency Region 10, 2001. Consensus plan for soil and sediment studies: Coeur d›Alene river soils and sediments bioavailability studies (URS DCN: 4162500.06161.05.a. EPA:16.2), pp. 1–16http://yosemite.epa.gov/R10/CLEANUP.NSF/fb6a4e3291f5d28388256d140051048b/503bcd6aa1bd60a288256cce00070286/$FILE/soil_amend_consensus_final_022801.PDF, 2012.

121. C. S. Reynolds and P. S. Davies, "Sources and bioavailability of phosphorus fractions in freshwaters: a British perspective," Biological Reviews of the Cambridge Philosophical Society, vol. 76, no. 1, pp. 27–64, 2001. · ·

122. J. N. Moore, W. H. Ficklin, and C. Johns, "Partitioning of arsenic and metals in reducing sulfidic sediments," Environmental Science and Technology, vol. 22, no. 4, pp. 432–437, 1988.

123. G. T. Ankley, D. M. Di Toro, D. J. Hansen, and W. J. Berry, "Technical basis and proposal for deriving sediment quality criteria for metals," Environmental Toxicology and Chemistry, vol. 15, no. 12, pp. 2056–2066, 1996.

124. J. Liu, K. T. Valsaraj, and R. D. Delaune, "Inhibition of mercury methylation by iron sulfides in an anoxic sediment," Environmental Engineering Science, vol. 26, no. 4, pp. 833–840, 2009. · ·

125. J. M. Benoit, C. C. Gilmour, R. P. Mason, and A. Heyes, "Sulfide controls on mercury speciation and bioavailability to methylating bacteria in sediment pore waters," Environmental Science and Technology, vol. 33, no. 6, pp. 951–957, 1999. · ·

126. A. Drott, L. Lambertsson, E. Bjorn, and U. Skyllberg, "Importance of dissolved neutral mercury sulfides for methyl mercury production in contaminated sediments," Environmental Science and Technology, vol. 41, no. 7, pp. 2270–2276, 2007. ·

127. Z. Xiong, F. He, D. Zhao, and M. O. Barnett, "Immobilization of mercury in sediment using stabilized iron sulfide nanoparticles," Water Research, vol. 43, no. 20, pp. 5171–5179, 2009. · ·

128. N. W. Revis, T. R. Osborne, G. Holdsworth, and C. Hadden, "Distribution of mercury species in soil from a mercury-contaminated site," Water, Air, and Soil Pollution, vol. 45, no. 1-2, pp. 105–113, 1989.

129. D. Renock, T. Gallegos, S. Utsunomiya, K. Hayes, R. C. Ewing, and U. Becker, "Chemical and structural characterization of As immobilization by nanoparticles of mackinawite (FeSm)," Chemical Geology, vol. 268, no. 1-2, pp. 116–125, 2009. · ·

130. M. Wolthers, L. Charlet, C. H. van Der Weijden, P. R. van der Linde, and D. Rickard, "Arsenic mobility in the ambient sulfidic environment: sorption of arsenic(V) and arsenic(III) onto disordered mackinawite," Geochimica et Cosmochimica Acta, vol. 69, no. 14, pp. 3483–3492, 2005. ·

131. T. J. Gallegos, P. H. Sung, and K. F. Hayes, "Spectroscopic investigation of the uptake of arsenite from solution by synthetic

mackinawite," Environmental Science and Technology, vol. 41, no. 22, pp. 7781–7786, 2007. · ·

132. T. J. Gallegos, Y. S. Han, and K. F. Hayes, "Model predictions of realgar precipitation by reaction of As(III) with synthetic mackinawite under anoxic conditions," Environmental Science and Technology, vol. 42, no. 24, pp. 9338–9343, 2008. · ·

133. Y. Liu, J. Terry, and S. Jurisson, "Pertechnetate immobilization with amorphous iron sulfide,"Radiochimica Acta, vol. 96, no. 12, pp. 823–833, 2008. · ·

134. R. R. Patterson, S. Fendorf, and M. Fendorf, "Reduction of hexavalent chromium by amorphous iron sulfide," Environmental Science and Technology, vol. 31, no. 7, pp. 2039–2044, 1997. ·

135. B. Hua and B. Deng, "Reductive immobilization of uranium(VI) by amorphous iron sulfide,"Environmental Science and Technology, vol. 42, no. 23, pp. 8703–8708, 2008. ·

136. E. C. Butler and K. F. Hayes, "Effects of solution composition and pH on the reductive dechlorination of hexachloroethane by iron sulfide," Environmental Science and Technology, vol. 32, no. 9, pp. 1276–1284, 1998. · ·

137. E. C. Butler and K. F. Hayes, "Kinetics of the transformation of trichloroethylene and tetrachloroethylene by iron sulfide," Environmental Science and Technology, vol. 33, no. 12, pp. 2021–2027, 1999. ·

138. E. C. Butler and K. F. Hayes, "Factors influencing rates and products in the transformation of trichloroethylene by iron sulfide and iron metal," Environmental Science and Technology, vol. 35, no. 19, pp. 3884–3891, 2001. · ·

139. X. Shi, K. Sun, L. P. Balogh, and J. R. Baker Jr., "Synthesis, characterization, and manipulation of dendrimer-stabilized iron sulfide nanoparticles," Nanotechnology, vol. 17, pp. 4554–4560, 2006.

140. C. Blodau, "A review of acidity generation and consumption in acidic coal mine lakes and their watersheds," Science of the Total Environment, vol. 369, no. 1–3, pp. 307–332, 2006. · ·

141. H. Niu and Y. Cai, "Adsorption and concentration of organic contaminants by carbon nanotubes from environmental samples," in Advances in Nanotechnology and the Environment, J. Kim,

Ed., pp. 79–136, Pan Stanford Publishing, Singapore, Singapore, 2012.

142. G. P. Rao, C. Lu, and F. Su, "Sorption of divalent metal ions from aqueous solution by carbon nanotubes: a review," Separation and Purification Technology, vol. 58, no. 1, pp. 224–231, 2007. · ·

143. Y. H. Li, J. Ding, Z. Luan et al., "Competitive adsorption of Pb2+, Cu2+ and Cd 2+ ions from aqueous solutions by multiwalled carbon nanotubes," Carbon, vol. 41, no. 14, pp. 2787–2792, 2003. · ·

144. C. Lu and C. Liu, "Removal of nickel(II) from aqueous solution by carbon nanotubes,"Journal of Chemical Technology and Biotechnology, vol. 81, no. 12, pp. 1932–1940, 2006. ·

145. H. Hyung, J. D. Fortner, J. B. Hughes, and J.-H. Kim, "Natural organic matter stabilizes carbon nanotubes in the aqueous phase," Environmental Science and Technology, vol. 41, no. 1, pp. 179–184, 2007. ·

146. D. P. Jaisi and M. Elimelech, "Single-walled carbon nanotubes exhibit limited transport in soil columns," Environmental Science and Technology, vol. 43, no. 24, pp. 9161–9166, 2009. · ·

147. D. P. Jaisi, N. B. Saleh, R. E. Blake, and M. Elimelech, "Transport of single-walled carbon nanotubes in porous media: filtration mechanisms and reversibility," Environmental Science and Technology, vol. 42, no. 22, pp. 8317–8323, 2008. · ·

148. L. Jiang, L. Gao, and J. Sun, "Production of aqueous colloidal dispersions of carbon nanotubes," Journal of Colloid and Interface Science, vol. 260, no. 1, pp. 89–94, 2003. · ·

149. M. J. O‹Connell, P. Boul, L. M. Ericson et al., "Reversible water-solubilization of single-walled carbon nanotubes by polymer wrapping," Chemical Physics Letters, vol. 342, no. 3-4, pp. 265–271, 2001. · ·

150. X. Zhou, L. Shu, H. Zhao et al., "Suspending multi-walled carbon nanotubes by humic acids from a peat soil," Environmental Science and Technology, vol. 46, no. 7, pp. 3891–3897, 2012. ·

151. L. Xueying, D. M. O‹Carroll, E. J. Petersen, H. Qingguo, and C. L. Anderson, "Mobility of multiwalled carbon nanotubes in porous media," Environmental Science and Technology, vol. 43, no. 21, pp. 8153–8158, 2009. ·

152. D. B. Warheit, B. R. Laurence, K. L. Reed, D. H. Roach, G. A. M. Reynolds, and T. R. Webb, "Comparative pulmonary toxicity assessment of single-wall carbon nanotubes in rats,"Toxicological Sciences, vol. 77, no. 1, pp. 117–125, 2004. · ·

153. C. W. Lam, J. T. James, R. McCluskey, and R. L. Hunter, "Pulmonary toxicity of single-wall carbon nanotubes in mice 7 and 90 days after intractracheal instillation," Toxicological Sciences, vol. 77, no. 1, pp. 126–134, 2004. · ·

154. G. Jia, H. Wang, L. Yan et al., "Cytotoxicity of carbon nanomaterials: single-wall nanotube, multi-wall nanotube, and fullerene," Environmental Science and Technology, vol. 39, no. 5, pp. 1378–1383, 2005. ·

155. A. Magrez, S. Kasas, V. Salicio et al., "Cellular toxicity of carbon-based nanomaterials," Nano Letters, vol. 6, no. 6, pp. 1121–1125, 2006. · ·

156. X. Chen, U. C. Tam, J. L. Czlapinski et al., "Interfacing carbon nanotubes with living cells,"Journal of the American Chemical Society, vol. 128, no. 19, pp. 6292–6293, 2006. · ·

157. X. Q. Li, D. G. Brown, and W. X. Zhang, "Stabilization of biosolids with nanoscale zero-valent iron (nZVI)," Journal of Nanoparticle Research, vol. 9, no. 2, pp. 233–243, 2007. · ·

158. N. G. Turan, "The effects of natural zeolite on salinity level of poultry litter compost,"Bioresource Technology, vol. 99, no. 7, pp. 2097–2101, 2008. · ·

159. A. A. Zorpas, T. Constantinides, A. G. Vlyssides, I. Haralambous, and M. Loizidou, "Heavy metal uptake by natural zeolite and metals partitioning in sewage sludge compost,"Bioresource Technology, vol. 72, no. 2, pp. 113–119, 2000. · ·

160. L. R. Nissen, N. W. Lepp, and R. Edwards, "Synthetic zeolites as amendments for sewage sludge-based compost," Chemosphere, vol. 41, no. 1-2, pp. 265–269, 2000. · ·

161. J. Villaseñor, L. Rodriguez, and F.J. Fernandez, "Composting domestic sewage sludge with natural zeolites in a rotary drum reactor," Bioresource Technology, vol. 102, no. 2, pp. 1447–1454, 2011. ·

162. A. A. Zorpas and M. Loizidou, "Sawdust and natural zeolite as a bulking agent for improving quality of a composting product

from anaerobically stabilized sewage sludge," Bioresource Technology, vol. 99, no. 16, pp. 7545–7552, 2008. ·

163. A. A. Zorpas, I. Vassilis, M. Loizidou, and H. Grigoropoulou, "Particle size effects on uptake of heavy metals from sewage sludge compost using natural zeolite clinoptilolite," Journal of Colloid and Interface Science, vol. 250, no. 1, pp. 1–4, 2002. · ·

164. A. A. Zorpas, A. G. Vlyssides, and M. Loizidou, "Dewatered anaerobically-stabilized primary sewage sludge composting: metal leachability and uptake by natural clinoptilolite," Communications in Soil Science and Plant Analysis, vol. 30, no. 11-12, pp. 1603–1613, 1999.

165. V. P. Gadepalle, S. K. Ouki, R. Van Herwijnen, and T. Hutchings, "Immobilization of heavy metals in soil using natural and waste materials for vegetation establishment on contaminated sites," Soil and Sediment Contamination, vol. 16, no. 2, pp. 233–251, 2007. · ·

166. H. Andry, T. Yamamoto, and M. Inoue, "Influence of artificial zeolite and hydrated lime amendments on the erodibility of an acidic soil," Communications in Soil Science and Plant Analysis, vol. 40, no. 7-8, pp. 1053–1072, 2009. · ·

167. T. Yamamoto, A. Yuya, A. Satoh, et al., "Application of artificial zeolite to combat soil erosion," in Proceedings of the American Society of Agricultural Engineers, Canadian Society for Engineering of Agricultural, Food and Biological System Annual International Meeting, Government Centre Ottawa, Ontario, Canada, August 2004.

168. M. Zheng, A technology for enhanced control of erosion, sediment and metal leaching at disturbed land using polyacrylamide and magnetite nanoparticles [M.S. thesis], Auburn University, Auburn, Ala, USA, 2011.

169. Z. S. Wang, M. T. Hung, and J. C. Liu, "Sludge conditioning by using alumina nanoparticles and polyelectrolyte," Water Science and Technology, vol. 56, no. 8, pp. 125–132, 2007. · ·

170. C. Ovenden and H. Xiao, "Flocculation behaviour and mechanisms of cationic inorganic microparticle/polymer systems," Colloids and Surfaces A, vol. 197, no. 1–3, pp. 225–234, 2002. · ·

171. Z. Yan and Y. Deng, "Cationic microparticle based flocculation and retention systems,"Chemical Engineering Journal, vol. 80, no. 1–3, pp. 31–36, 2000. · ·

172. T. H. Eyde, "Zeolites," Minerals Engineering, vol. 62, p. 86, 2010.

173. X. Q. Li, D. W. Elliott, and W. X. Zhang, "Zero-valent iron nanoparticles for abatement of environmental pollutants: materials and engineering aspects," Critical Reviews in Solid State and Materials Sciences, vol. 31, no. 4, pp. 111–122, 2006.

174. Carbon nanotube, http://en.wikipedia.org/wiki/Carbon_nanotube, 2012.

Chapter 6

Numerical Study on an Applicable Underground Mining Method for Soft Extra-Thick Coal Seams in Thailand

Nay Zarlin, Takashi Sasaoka, Hideki Shimada, and Kikuo Matsui

Department of Earth Resources Engineering, Kyushu University, Fukuoka City, Japan

ABSTRACT

The EGAT Mae Moh Mine is the largest open pit lignite mine in Thailand and it produces lignite about 16 million tons annually. In the near future, the pit limit of the mine will be reached and underground mine will then be developed through the open pit in the depth of 400 - 600 m from the surface. However, due to the challenges for underground mining such as poor geological conditions, extra thickness (20 - 30 m) of coal seams, and weak mechanical properties of coal seams and the surrounding rock, the success possibility of underground mining and an applicable underground mining method is being investigated at the present. The paper discusses the applicability of multi-slice bord-and-pillar method for the soft extra thick coal seams in the Mae Moh mine by means of numerical analyses using the 3D finite difference code "FLAC3D".

INTRODUCTION

Thick coal seams have been mined by various methods, such as extended height single pass longwall method, longwall top coal caving method, slicing method, blasting gallery method, and hydraulic mining method etc., depending on local geological and geotechnical conditions and thickness, depth and characteristics of seams. Thick seam mining is very different from conventional coal mining in many aspects. It is rather complicated to predict the characteristics of strata response to mining operation and the mining of such seams is beset with numerous problems [1-3]. Although there have been many research works that focused on the thick seams in medium to strong geotechnical conditions, very few researches and publications are found for the extra-thick seams that are over 10 m in thickness in weak conditions. In this study, it investigated the ground response to mining and failure mechanism in soft extra-thick seams by means of numerical analyses and discussed an applicable underground method for mining the soft extra-thick coal seams in Mae Moh mine in Thailand.

BRIEF INFORMATION OF MAE MOH MINE

General

Mae Moh mine is located in Mae Moh district, Lampang province, about 630 kilometers north of Bangkok, in Thailand. The location map of Mae Moh mine is shown in Figure 1. This mine is operated by EGAT (Electricity Generating Authority of Thailand), and it is the largest open-pit lignite mine in Thailand. The total geological and economical lignite reserves are approximately 1140 million tons and 825 million tons, respectively. The annual production is about 16 million tons, which represents 70% of the total coal production of Thailand. All of the lignite produced from the mine is supplied to the 2400 MW Mae Moh power plant that is providing 15% of the total electricity demand of Thailand. About 347 MT of lignite has been produced, and the remaining future reserve is approximately 478 MT as of 2011.

The current Mae Moh pit covers an area of 4 km by 7.5 km at various depths up to 290 m as of 2011. According to the mining plan reported by the EGAT, the final pit limit will be reached and the open pit operation will be finished in the near future. After that, underground mining will be commenced through the final pit high wall from the depth of about 400 m. The total lignite reserves in the underground mine area are approximated about 150 million tons (see Figure 2). However, due to the adverse geological and geotechnical conditions such as extra thickness of coal seams, and weak mechanical properties of coal seams and the surrounding rock, success possibility of underground mining and an applicable underground mining method for Mae Moh mine is being investigated at the present [4,5].

Geologic Settings of Mae Moh Mine

Mae Moh mine is situated in Mae Moh tertiary basin which is more than 1000 m of maximum thickness. Three main geological formations, namely Huai Luang (HL), Na Khaem (NK) and Huai King (HK), are found in the Mae Moh basin (see Figure 3).

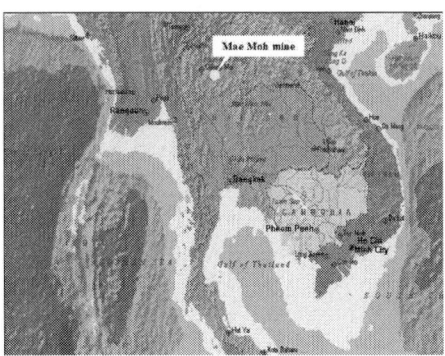

Figure 1: Location of Mae Moh mine.

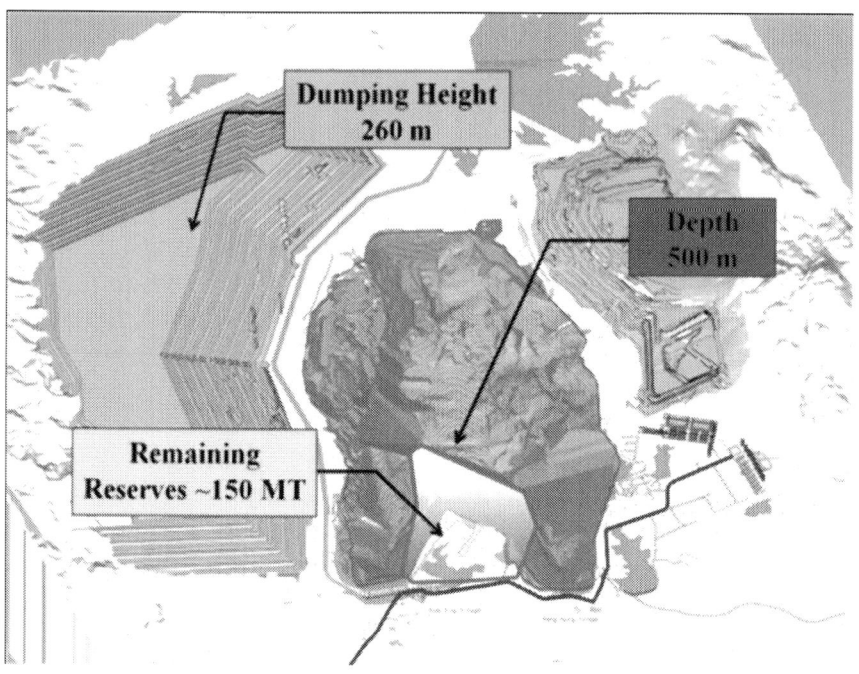

Figure 2: Final pit of Mae Moh mine (plan).

HL formation mainly consists of red to brownish-red semi-consolidated and unconsolidated clay, silt and sandstone. NK formation composes lignite seams and gray to greenish-gray claystone and mudstone, whereas HK formation consists of semi-consolidated fine to coarse sand-stone, claystone, mudstone and conglomerate with green, yellow, blue and purple in color. Five lignite seams, marked as J, K, Q, R and S seam, are found in the NK formation. However, the J, R and S seams are considered uneconomical due to the poor quality/ thickness and depth of seam, and thus major economical mineable seams are only K and Q seams. The thickness of K and Q seams ranges from 20 to 30 m with the interburden of 20 to 25 m mudstone, up to 600 m depth from the surface [4,5].

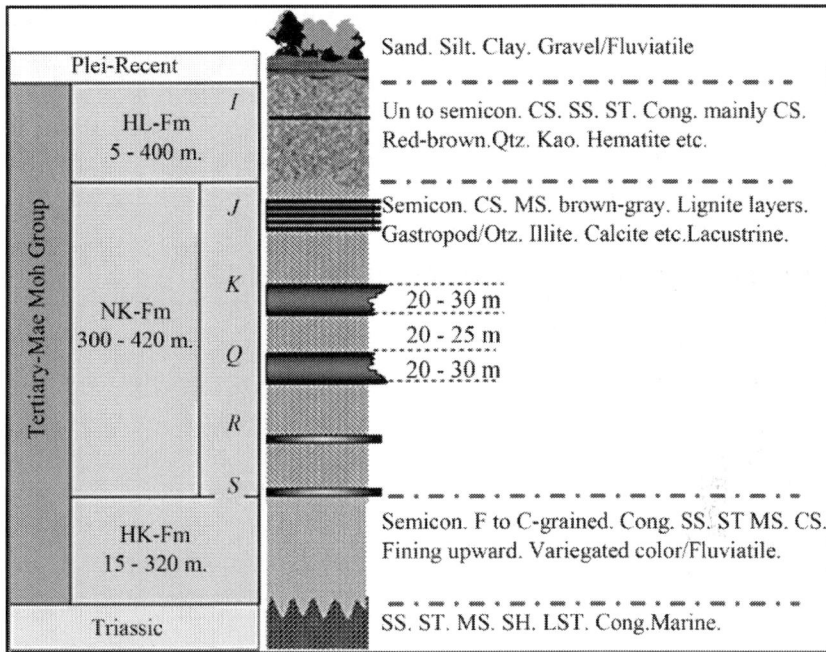

Figure 3: Generalized stratigraphic column.

NUMERICAL MODELING

In underground mining of a coal seam that exceeds 10 m in thickness, extraction is mostly considered as dividing the seam into a number of slices/lifts and extracting each slice/lift with a longwall or bord-and-pillar using continuous or conventional mining methods. In a multi-slice mining, the longwall caving system is the preferred practice; but, the material produced requires a large expense to compensate for surface/subsurface damage. Furthermore, it is possible for nearby seas, lakes, rivers, creeks, canals, and other surface features to be disturbed if serious damage occurs [6,7]. For the Mae Moh mine, however, the application of longwall method in underground mining is considered limited potential since the properties of rocks are very weak and the coal seams are extra-thick, that might occur large ground disturbance/ subsidence and thus it is possible to occur the pit instability and many problems. Therefore, the multi-slice bord-andpillar mining was firstly considered in this study and its applicability in soft extra-thick seams was investigated by means of numerical analyses using three dimensional finite difference program Flac3D.

Description of the Numerical Model

The Flac3D numerical model was composed of 20 m and 25 m thick double coal seams, named as k_seam and q_seam, with the interburden of 25 m and overburden of 400 m in thickness. For simplicity, multi thin layers were not considered and the overburden and interburden were modeled as homogeneous mudstone layers. Due to the limitations of computer running time and capacity restriction, only the upper coal seam, k_seam, was modelled for extraction and set up with a smaller mesh size in order to obtain more precise stress, distribution ground displacement, and failure distribution around the extraction area. The total dimension of the model is 220 m wide, 220 m long and 470 m high, respectively. The numerical model consists of 448,000 elements and 468,671 grid points. The geometry, density of grids and group of materials of the model are presented in Figure 4. The bottom of the model was restricted in the vertical direction, whereas the sides of the model were restrained perpendicular to each side. Mohr Coulomb elasto-plastic constitutive relation was employed in the analyses. The mechanical properties of the materials used in the numerical analyses are presented in Table 1. After defining the constitutive relation and material properties, and assigning boundary conditions and initial state, the model was run until the equilibrium stage had been reached. After the response of the model was satisfactory, formulation of bord-and-pillar mine was conducted by excavation of galleries and formation of pillars sequentially.

Results and Discussion

At first, the ground response and pillar conditions in different pillar sizes were investigated in order to determine an appropriate pillar size. The width and length of the panel was taken as 100 m in the X direction and 150 m in Y direction. The dimension of each gallery was taken as 6 m in width and 3 m in height, respectively. In the case of Mae Moh coal field where coal seams are soft and extra-thick, multi slicing with ascending order of extraction is considered limited potential as ascending order does not allow caving into the galleries [7].

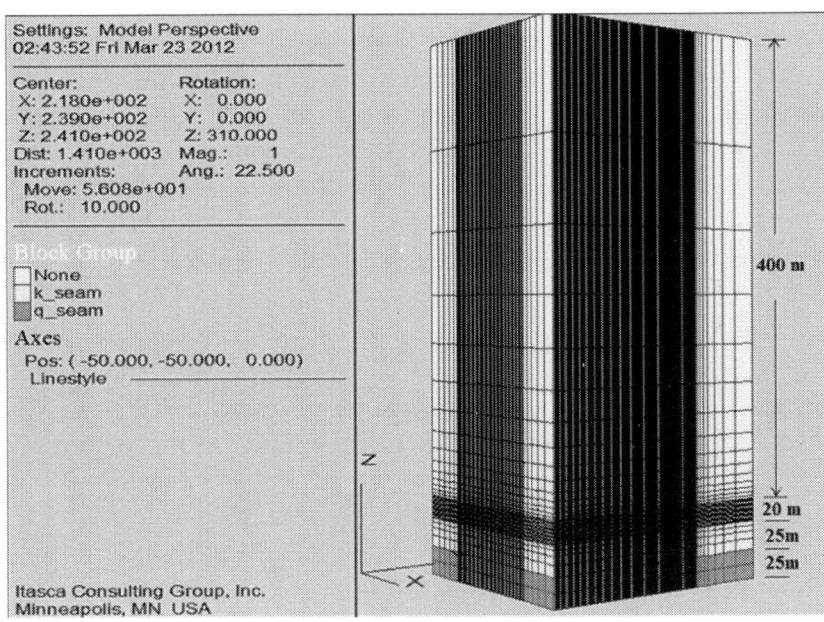

Figure 4: Flac3D numerical model.

Table 1: Mechanical properties of the materials used in the analyses

Parameters	Coal	Mudstone
Density (kg/m³)	1430	1950
Bulk modulus (MPa)	1250	1400
Shear modulus (MPa)	576.92	840
Friction angle (°)	22.3	33.5

Cohesion (MPa)	0.8	1.2
Tensile strength (MPa)	0.3	1

Therefore, the first slice was modeled along the roof and the next slices were then modeled sequentially in descending order. Figures 5 and 6 show the failure states at different size of pillars after the first slice was developed along the mine roof. The pillar width to height ratio (w/h) is 4 in the Figures 5 and 6 in the Figure 6, respectively. In Flac3D, the terms in legend of figures showing block state "none" indicate no-failure zone, "shear-n" and "tension-n" indicate yield in shear and tension now, "shear-p" and "tension-p" indicate elastic state now, but yield in shear and tension past, respectively [8].

According to the results, it could be seen that all the pillars in the Figure 5 failed (shear-n, tension-n) whereas pillars seemed to be in stable in Figure 6 although partial failures occurred. Therefore, the pillars size was considered starting from the w/h of 6 in the analyses. Figures 7 and 8 show the conditions of pillars, with the w/h of 6, after the second slice was developed in superimposed and non-superimposed pattern, where the coal parting 3 m was left between the slices. According to these results, it was expected pillar instability and roof control problems in both patterns of developments; all pillars and parting failed in superimposed pattern development whereas pillars at the first slice and parting failed in nonsuperimposed pattern development. Subsequently, the pillars size and parting thickness were increased and the response was investigated. However, it was found that the conventional multi-slice bord-and-pillar could not be successful in the soft extra-thick seams due to the problems in roof control and failures of partings and openings when the lower slices were extracted.

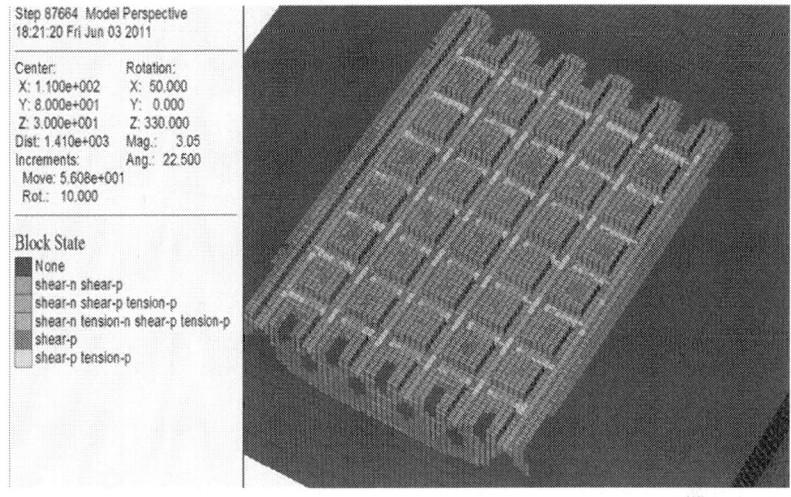

Figure 5: Failure state at pillars after developing first slice along the roof (pillar width of 12 m (w/h = 4)).

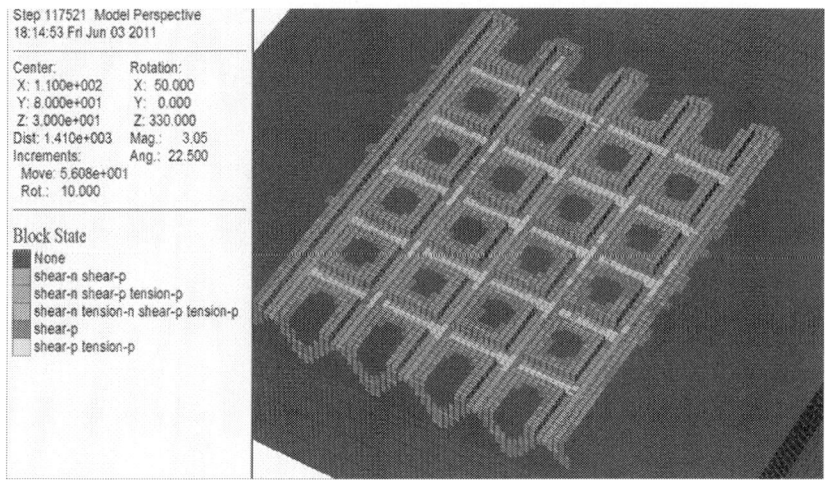

Figure 6: Failure state at pillars after developing first slice along the roof (pillar width of 18 m (w/h = 6)).

Due to the situation discussed above, it was investigated the performance of an alternative method in which stowing and different cutting method were employed. The purpose of stowing is not to transmit the rock stresses, but to reduce the relaxation of the rock mass so the rock itself will retain a load carrying capacity and will improve load shedding to crown pillars and abutments. This leads to less deterioration in ground conditions in the mine, improving operations and safety [9]. The required strength of the stowing materials depends on the strata and mining conditions: such as cover depth, rock type and properties, mining method, etc. In a shallow mine, the required strength is not as critical when compared to that for a deep mine. Considering the rock properties at the vicinity of the production area and the depth of the seams at the mine, the strength of stowing material was initially taken as one half of the coal strength in the study [10-12]. The alternative coal cutting method and stowing were carried out as shown in Figure 9.

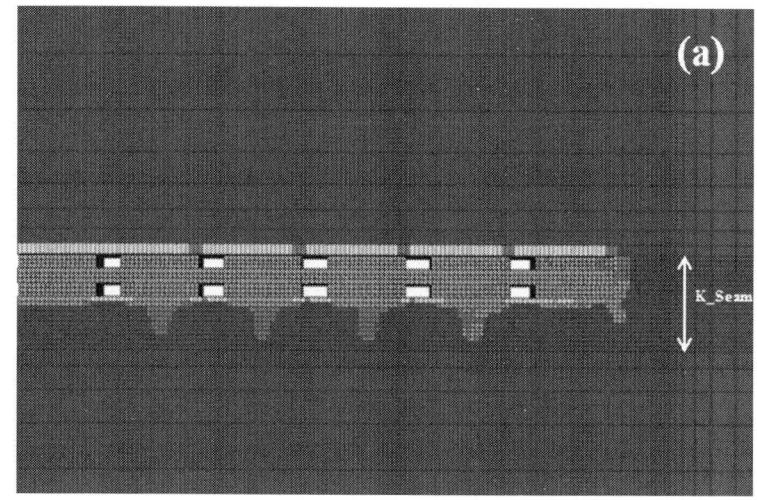

(a)

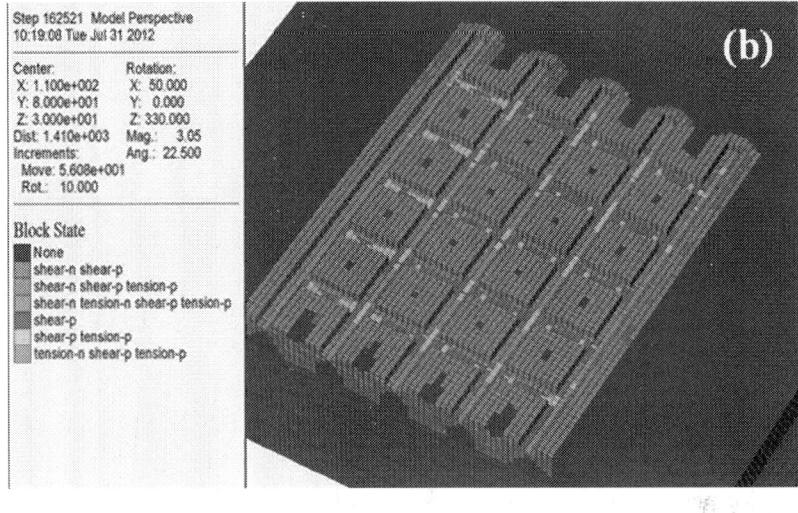

(b)

Figure 7: Failure state after developing the second slice in superimpose d pattern (a) cutting view along the central part of the panel and (b) pillars in the second slice.

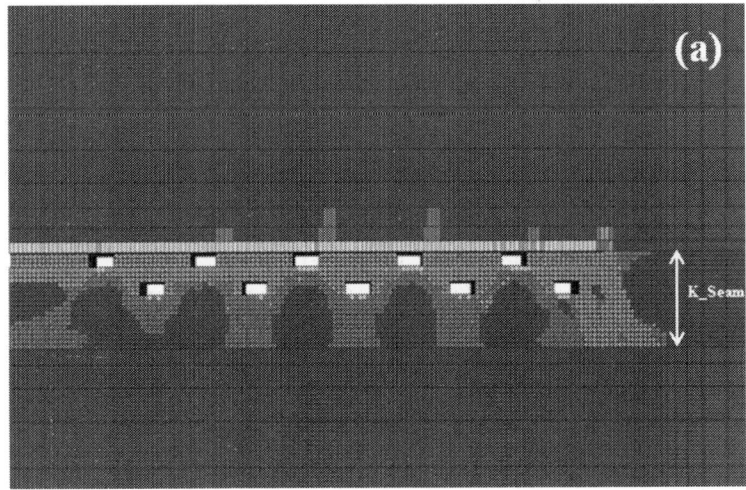

(a)

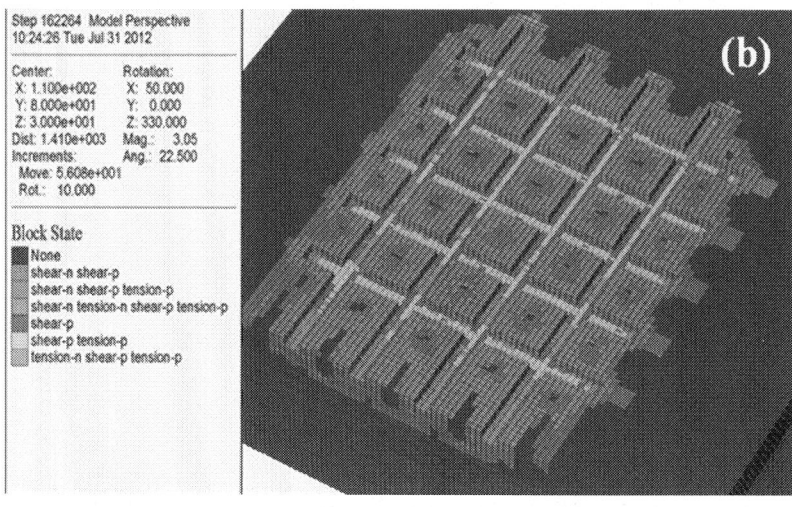

(b)

Figure 8: Failure state after developing the second slice in non-superimposed pattern (a) cutting view along the central part of the panel and (b) pillars in the second slice.

First, pairs of entries were cut across the mine roof by leaving 18 m wide pillars between them (see Figure 9(a)). The width of each entry was also taken as 6 m and mining height was 3 m. After that the cut entries were stowed (see Figure 9(b)) and a pairs of entries were then cut next to the stowed entries (see Figure 9(c)). The mining and stowing were repeated in the same manner until the whole panel was extracted (see Figure 9(d)). According to the results of failure state shown in Figures 9(a)-(d), it was found that the pillars maintained their stability until the whole panel was extracted. The confinement produced by stowing improved the pillars' stability throughout the panel extraction. Figures 10(a) and (b) show contours of vertical displacement and failure state after cutting pairs of entries at the second slice. According to the results, it was found a very few failures zones (shear-n, tension-n) at the roof and small displacement about 2 - 4 cm in maximum around the openings and therefore, the problems in roof control and pillars failure cannot be expected. Mining and stowing were repeated in the same manner as the first slice. The next slices extractions were also done slice by slice sequentially in descending order and no large roof control and pillars instability problems were expected during the extractions. Figures 11(a) and (b) show contours of vertical displacement and failure state after extracting the whole k_seam. It could be seen that the ground displacement and failures around the extracted seam are very small even after the whole seam was extracted. Therefore, it could be supposed that this is an applicable and effective method for mining soft extra-thick seams in terms of diminishing ground disturbance, mine safety and maximizing coal recovery.

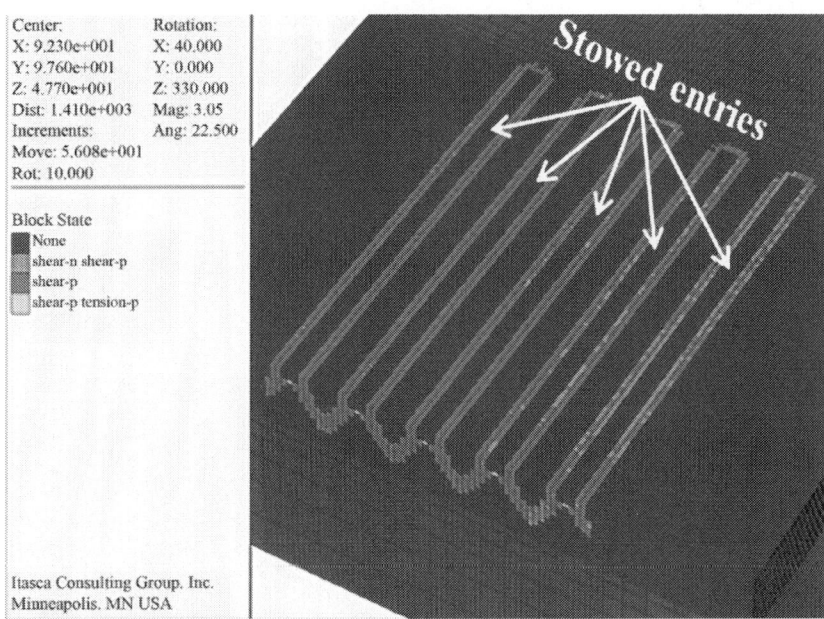

(a)

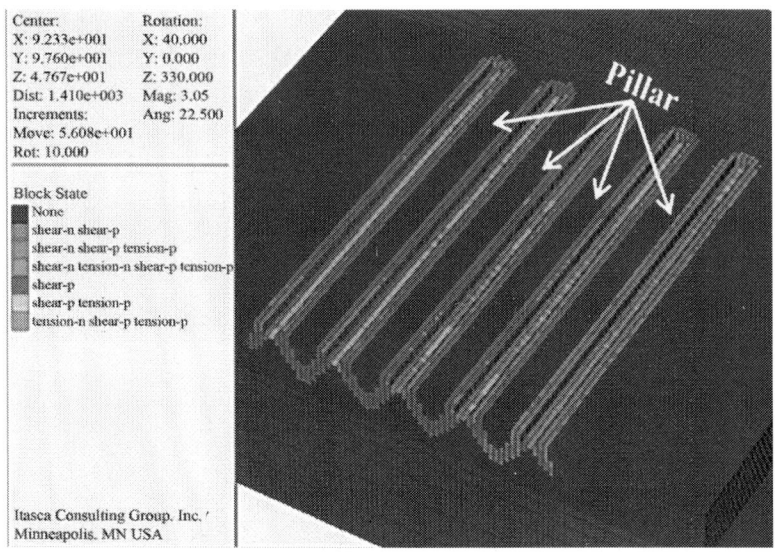

(b)

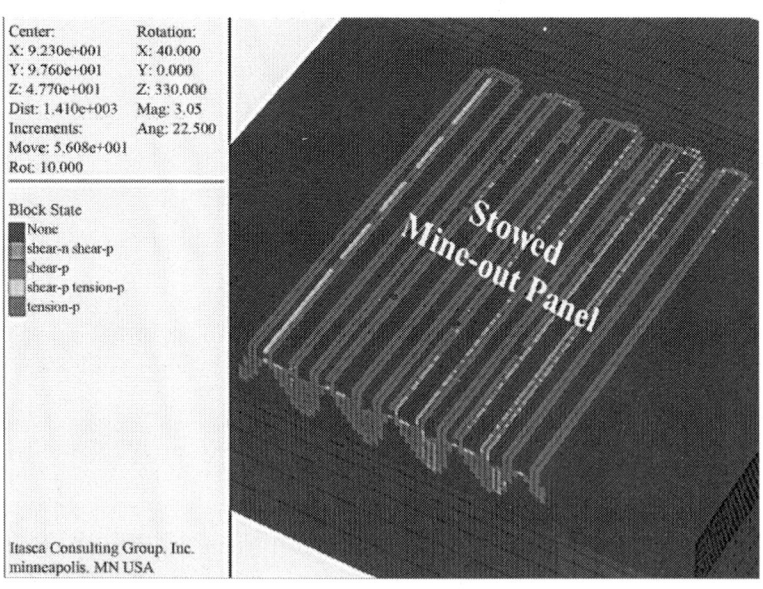

(c)

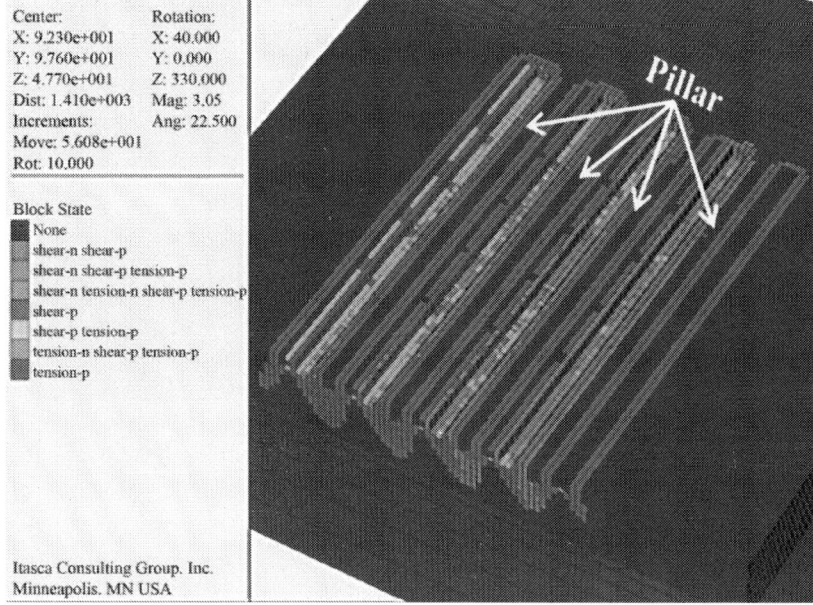

Center: Rotation:
X: 9.230e+001   X: 40.000
Y: 9.760e+001   Y: 0.000
Z: 4.770e+001   Z: 330.000
Dist: 1.410e+003   Mag: 3.05
Increments: Ang: 22.500
Move: 5.608e+001
Rot: 10.000

Block State
None
shear-n shear-p
shear-n shear-p tension-p
shear-n tension-n shear-p tension-p
shear-p
shear-p tension-p
tension-n shear-p tension-p
tension-p

Itasca Consulting Group. Inc.
Minneapolis. MN USA

(d)

Figure 9: Failure states along with the sequences of mining and stowing at the first slice (a) after 1st cut; (b) after stowing the 1st cut; (c) after 2nd cut and (d) after extracting and stowing the whole panel.

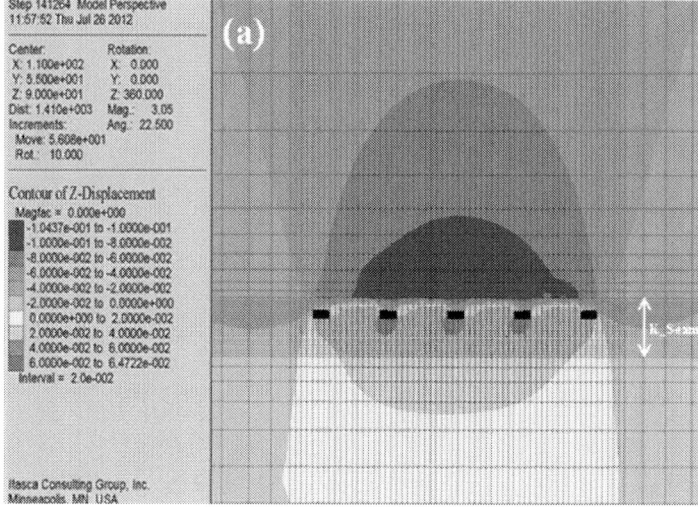

(a)

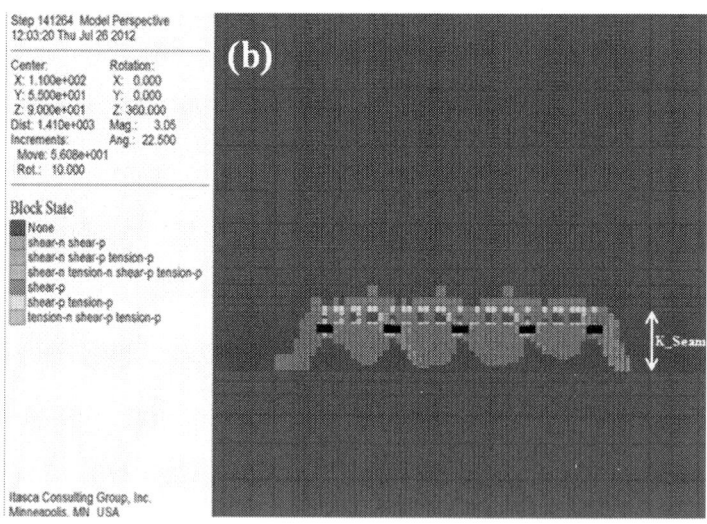

(b)

Figure 10: (a) Vertical displacement contours and (b) failure state after 1ˢᵗ cut at the second slice.

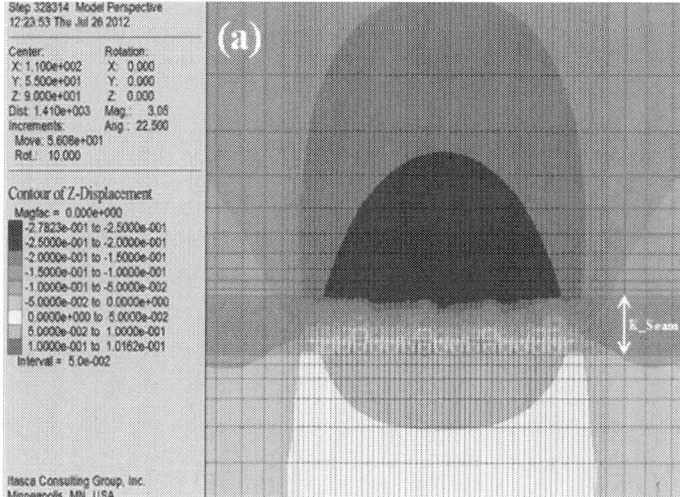

(a)

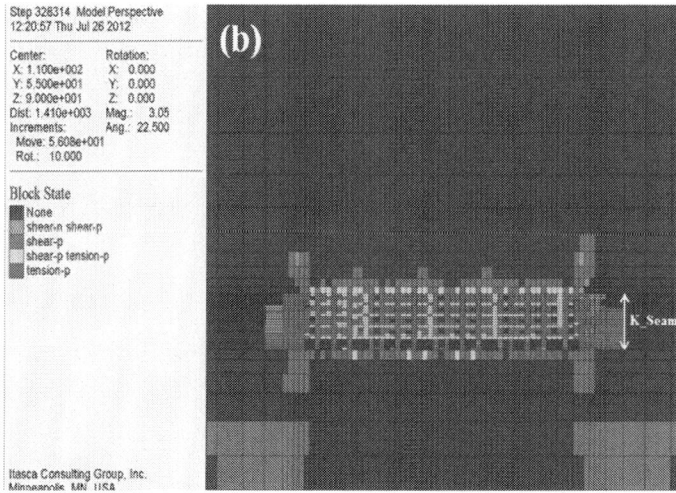

(b)

Figure 11: (a) Vertical displacement contours and (b) failure state after extracting the k_seam and stowing in the mine-out voids.

CONCLUSIONS

According to the results of a series of numerical analysis, it was found that multi-slice bord-and-pillar mining with the alternative method of cutting and stowing discussed in this paper can be employed for the soft extra-thick coal seams and this is an effective method to diminish ground disturbance/environmental impacts, to improve mine safety and to obtain maximum coal recovery. However, it must be evaluated more details for stowing, such as the required strength of stowing material, the method of stowing to be employed as well as the cost and economy of mine, before employing this system.

ACKNOWLEDGEMENTS

The authors wish to express their gratitude to the managers, engineers and miners of EGAT Mae Moh lignite mine for arranging their mine site visit and for providing information. All opinion and comments stated in this paper are those of authors and do not necessarily represent those of the institutions or the mine.

REFERENCES

1.   R. D. Singh, "Principles and Practices of Modern Coal Mining," New Age International (P) Ltd., New Delhi, 1997, pp. 249-285.

2.   M. K. Ozfirat, F. Simsir and A. Gonen, "A Brief Comparison of Longwall Methods Used at Mining of Thick Coal Seams," Proceedings of the 19th International Mining Congress and Fair of Turkey, Turkey, 9-12 June 2005, pp. 141-144.

3.   H. Furukawa, K. Matsui and H. Shimada, "The Present Situation and Issues in Coal Mining in India," Journal of the Mining and Materials Processing Institute of Japan, Vol. 121, No. 9, 2005, pp. 446-455. doi:10.2473/shigentosozai.121.446

4.   EGAT, "Mae Moh Mine Geotechnical and Slope Stability," Powerpoint Slides and Reports, 2011.

5.   P. Doncommul, P. Pimklang, N. Mungpayabal and A. Chaiwan, "Reviews of Slope Failure in Lowwall Area 3 of Mae Moh Mine,"

Fuenkajorn and Phienwei, Eds., Rock Mechanics, 2011, pp. 219-227.

6.   K. Matsui, H. Shimada, T. Sasaoka and H. Furukawa, "Some Consideration in Underground Mining Systems for Extra-Thick Coal Seam," Coal International, Vol. 259, No. 2, 2011, pp. 38-41.

7.   M. L. Jeremic, "Strata Mechanics in Coal Mining," A. A. Balkema Publishers, Leiden, 1985.

8.   Itasca Consulting Group, Inc., "Flac3D Command Reference," Mill Place, Minneapolis, 2003, p. 141.

9.   J. R. Barrett, M. A. Coulthard and R. M. Dight, "Determination of Fill Stability," Mining with Backfill—12th Canadian Rock Mechanics Symposium, CIM Special Vol. 19, Sudbury, 23-25 May, 1978, pp. 85-91.

10.   D. R. Tesarik, J. B. Seymour, T. R. Yanske and R. W. Mckibbin, "Stability Analysis of a Backfilled Room-andPillar Mine," Report of Investigations, United States Bureau of Mines, 1995, pp. 1-26.

11.   T. Grice, "Underground Mining with Backfill," The 2nd Annual Summit-Mine Tailings Disposal System, Brisbane, 24-25 November 1998, pp. 1-14.

12.   K. Matsui, H. Shimada, S. Kramadibrata and M. Zrai, "Some Considerations of Highwall Mining Systems in Coal Mines," Proceedings of the 17th International Mining Congress and Exhibition of Turkey, Ankara, 19-22 June 2001, pp. 269-276.

Chapter 7

Vegetation of Mono-Layer Landfill Cover Made of Coal Bottom Ash and Soil by Compost Application[*]

Seul Bi Lee[1], Sang Yoon Kim[2], Chan Yu[3], Soon-Oh Kim[4], and Pil Joo Kim[2, 5]

[1]Department of Agricultural Environment, National Academy of Agricultural Science (NAAS), Rural Development Administration (RDA), Suwon, South Korea

[2]Division of Applied Life Science (BK 21 Program), Gyeongsang National University, Jinju, South Korea

[3]Department of Agricultural Engineering, Gyeongsang National University, Jinju, South Korea

[4]Department of Earth and Environmental Sciences, Gyeongsang National University, Jinju, South Korea

[5]Institute of Agriculture and Life Science, Gyeongsang National University, Jinju, South Korea;

ABSTRACT

Monolayer barriers called evapotranspiration (ET) covers were developed as alternative final cover systems in waste landfills but high-quality soil remains a limiting factor in these cover systems. Coal bottom ash was evaluated to be a very good alternative to soil in previous tests and a combination of soil (65% wt·wt^{-1}) and coal bottom ash (35% wt·wt^{-1}) was evaluated to be the most feasible materials for ET cover systems. In our pot test, selected manure compost as soil amendment for the composite ET cover system, which was made of soil and bottom ash at ca. 40 Mg·ha^{-1} application level was very effective to promote vegetation growth of three plants; namely, garden cosmos (Cosmos bipinnatus), Chinese bushclover (Lespedeza cuneata), and leafy lespedeza (Lespedeza cyrtobotrya). To evaluate the effect of compost application on plant growth in an ET vegetative cover system, two couples of lysimeters, packed with soil and a mixture of soil and bottom ash, were installed in a pilot landfill cover system in 2007. Manure composts were applied at the rates of 0 and 40 Mg·ha^{-1} before sowing the five plant species, i.e. indigo-bush (Amorpha fruticosa), Japanese mugwort (Artemisia princeps, Arundinella hirta, Lespedeza cuneata, and Lespedeza cyrtobotrya). Unseeded native plant (green foxtail, Setaria viridis) was dominant in all treatments in the 1st year after installation while the growth of the sown plants significantly improved over the years. Total biomass productivity significantly increased with manure compost application, and more significantly increased in the composite ET cover made of soil and bottom ash treatment compared to the single soil ET cover, mainly due to more improved soil nutrient levels promoting vegetation growth and maintaining the vegetation system. The use of bottom ash as a mixing material in ET cover systems has a strong potential as an alternative to fine-grained soils, and manure compost addition can effectively enhance vegetative propagation in ET cover systems.

INTRODUCTION

Landfills undergoing closure must be covered with a final cover that minimizes the long-term migration of liquids through the landfill [1]. The capping system can vary from simple soil cover to multiple layers of earthen and geosynthetic materials [2]. Several studies [3,4] have explored various alternative cover technologies for final closure of waste landfills. Among them, monolayer barriers called as evapotranspiration (ET) cover are covers that include a thick layer of fine-grained soil generally covered with a layer of vegetated topsoil and alternative final cover systems to the conventional cover system.

Different to conventional cover system designs that use materials with low hydraulic permeability, ET cover systems use water balance components to minimize percolation. These cover systems rely on the properties of soil to store water until it is either transpired through vegetation or evaporated from the soil surface. This type of thick cover encourages water storage and enhances ET yearround, rather than just during the growing seasons. The soil allows water storage, which, when combined with the vegetation, will increase ET. These soil barriers can be cost effective when large quantities of fine grained soil requiring little processing is available on site. However, most of landfill sites in the world are struggling to find large amounts of good quality soil.

The materials used in soil-based cover systems are either natural materials, modified soils, synthetic material, or waste materials. Well-graded fine-grained compacted soils are usually selected in case of natural soils. If available, different types of clay are the most likely choice because of their low hydraulic conductivity and adequate performance in eliminating the fluids transport through landfills. There has been growing interest in using waste materials as alternative hydraulic barriers for conventional materials in lining and covering landfills. This is apparent where clay and other fine soils are not readily available and usually require high prices for transport from remote locations. Another reason is attributed to the huge amounts of generated wastes and the elevating costs associated with their disposal [5].

Among the waste materials that have already been used as substitute for soil-based covers are fly ash, slags from iron and steel-making, non-ferrous slags, domestic refuse incinerator ash, overburden materials, dredged silts, construction rubble, wastewater treatment sludges, and paper mill sludges. Mollamahmutoglu and Yilmaz [6] found that 20% bentonite-class F fly ash was suitable as a liner or cover material at waste disposal areas, and Kim et al. [7] found in the lab test that coal bottom ash among four industrial byproducts (blast furnace and steel refining slags, coal bottom ash, and phospho-gypsum) was the most feasible alternative of soil in the ET cover system and a mixture of ca. 35% of bottom with soil was the most suitable [8]. Bottom ash has a particle size generally within the range of 0.1 - 10 mm [9]. The chemical constituents of bottom ash can vary greatly depending on the coal type, source, and plant operating parameters. Major constituents include calcium (Ca), aluminum (Al), iron (Fe), magnesium (Mg), potassium (K), silicone (Si), sodium (Na) and titanium (Ti). These constituents typically constitute up to 95% of the mass of the ash. Of these materials, Ca, Fe, Mg, K and Si are essential plant nutrients [9].

However, fast re-vegetation could be of particular importance to efficiently establish ET cover systems, especially in the case of the industrial byproduct utilization, since leachate volume can decrease as a result of the soil-plant systems ET [10]. Plants within a soil-plant system can evapotranspirate a large amount of incoming water, including landfill leachate. Plant ET potential closely depends on plant growth, and therefore, soil fertility management can be very important. Manure compost application could be a simple and good amendment to improve soil fertility and plant growth. The benefits of using manure compost as an organic soil amendment may be seen in agricultural land. Recently animal wastes represent a disposal problem while offering potential soil amendment benefits in most countries. Since cattle feeding industry is continuously expanding in Korea, manure compost utilization in landfill cover as an amendment could be a good disposal area of manure.

In this view, the objective of this study was to determine the optimum application levels of manure based compost as a soil amendment in the ET cover system, which was developed by mixing bottom ash (35%, wt·wt^{-1}) and soil (65%, wt·wt^{-1}) [7,8], and then evaluate the effect of compost application on vegetation development and soil properties.

MATERIALS AND METHODS

Selection of Bottom Ash and Soil

In previous studies [7,8], the bottom ash among four industrial byproducts (blast furnace slag and steel refining slag from iron making factories, coal bottom ash for electric power station, and phospho-gypsum from chemical fertilizer factory) was selected as the best mixing material with soil for installing an ET cover system. In this test, the same bottom ash and soil were selected in the pot test and pilot landfill cover system, with the purpose of determining the optimum compost application rate and its field applicability, respectively.

The coal bottom ash was collected from a thermal power plant in Hadong Power Plant of Kwangyang, South Korea and air-dried and sieved to <4 mm for the pot and pilot tests. Characteristics of coal bottom ash were alkaline (pH 8.9) and had high concentration of available phosphorus. The soil that was collected from an alpine area in Gyeongsang National University campus, Jinju City, South Korea campus had a pH of 6.1 with low nutrient contents (Table 1).

Preparation of Pot Test

To determine the effect of compost application on the vegetative growth of three selected plants, namely, garden cosmos (Cosmos bipinnatus), Chinese bushclover (Lespedeza cuneata), and leafy lespedeza (Lespedeza cyrtobotrya), which are generally grown in landfill cover plantations in Korea, a horticultural seedling bed tray (L 80 cm × W 60 cm × D 20 cm size) was filled with the bottom ash (35%) and soil (65%) mixture. Four levels of compost (0, 20, 40, and 80 $Mg \cdot ha^{-1}$) were applied on the surface and totally hand-mixed. The compost material was purchased from a local market with typical characteristics of a swine manure compost (pH 6.8, organic matter 406 $g \cdot kg^{-1}$, total N 11 $g \cdot kg^{-1}$, C/N ratio 24, total P_2O_5 19 $g \cdot kg^{-1}$, and total K_2O 13 $g \cdot kg^{-1}$). Thirty seeds of each plant were seeded in a line of two rows in a seedling bed with constant intervals (10 cm × 5 cm) on April 9, 2006, grown under ambient conditions in a greenhouse, and harvested on November 20, 2006 for evaluating the total plant biomass. Moisture contents were controlled during the plant cultivation period following the conventional method for upland plants recommended by NAIST of Korea [11]. The treatments were replicated with three times.

Table 1: Chemical properties of soil and coal bottom ash used in the pot and vegetative cover pilot tests

Parameters	Soil	Coal bottom ash
pH (1:5 with H2O)	6.1	8.9
Electrical conductivity (dS.m-1)	0.22	1.4
Organic matter (g.kg-1)	14.5	33.1
Available P2O5 (mg-kg-1)	6.1	551
Exchangeable cations (cmol+ kg-1)		
K	0.12	0.03
Ca	4.3	4.9
Mg	2.6	1.60
Na	0.3	0.31
Soil Texture	Silt Loam(SiL)	–

Installation of the Vegetative Cover Pilot System

A vegetation test was conducted on a pilot scale using the lysimeter method. Four sets of lysimeter, each set with a dimension of H 1.2 m × W 2 m × L 6 m size, were constructed on the campus of Gyeongsang National University, Jinju, South Korea (Figure 1). This study was carried out in a typical monsoonal climate within a temperate zone and the annual mean temperature and precipitation were recorded to be 13.1°C and 1513 mm, respectively, over a 30-year period (1980-2010) [12].

A piezometer constructed of PCV tubes, each 5cm in diameter, was put into the lysimeter for ground water sampling, accumulation and level control. The piezometer was closed from the top and filtered from the bottom in a gravel layer. The gravel layer with a particle diameter of 10 to 20 mm allowed drainage of percolating water.

The four sets of lysimeter were packed with a mixture of soil (65%) and bottom ash (35%, two sets) and the pure soil (two sets). In the pot test, ca. 40 Mg·ha^{-1} of manure compost was evaluated as the optimum level in this ET cover condition. To determine the effect of compost application on re-vegetation, compost was applied with rates of 0 and 40 Mg·ha^{-1}, and then mixed manually at 20 cm depth. Selected plant seeds (Amorpha fruticosa, Artemisia princeps, Arundinella hirta, Lespedeza cuneata, Lespedeza cyrtobotrya) were broadcasted evenly and covered with a thin layer of soil in all lysimeters in the mid May, 2007. Thereafter, the vegetation was maintained without any further fertilizer or tillage activities and the above-ground parts of the plants were harvested around the end of October in 2007 and 2009, air-dried, and weighed for total biomass productivity.

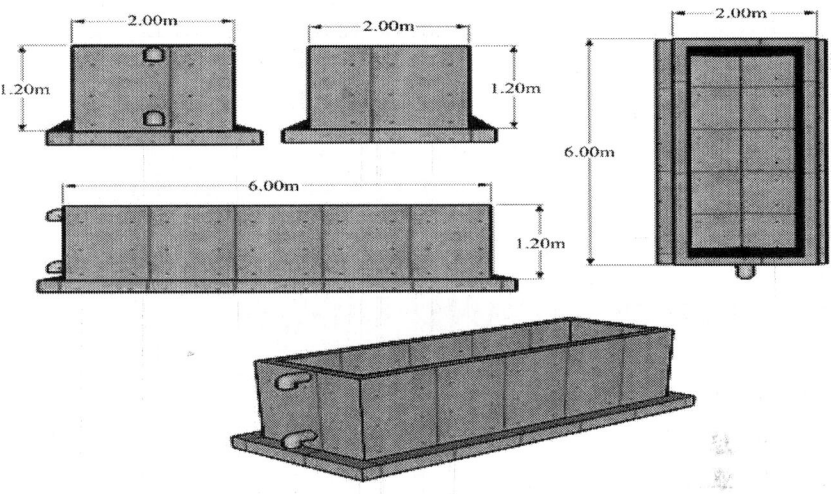

Figure 1: Layout of the ET landfill lysimeter chamber used in the vegetative cover pilot test.

Investigation of Plant Biomass Productivity and Soil Chemical Properties

The vegetative biomass was harvested in 0.5 m × 1.0 m size around the late October in the 1st and 3rd years after the installation (2007). The harvested plant biomass was oven-dried at 70°C for 72 hr, and then weighed on dry weight basis, which was replicated three times.

The compost used in the pot and pilot tests were ovendried at 70°C for 72 hr, ground and then digested using a ternary solution

$(HNO_3:H_2SO_4:HClO_4, 10:1:4$ volume·volume$^{-1})$ to determine the total P and K contents. Total C and N concentrations were quantified by CHNS Analyzer (CHNS-932 Analyzer, Leco, USA).

Surface soil samples (0 - 15 cm) were collected from the pilot system at the plant biomass harvesting stage in the 1st and 3rd years after the installation, air-dried and sieved (<2 mm) for chemical analysis. The chemical properties were analyzed as follows: pH (1:5 water extraction), organic matter content (Walkley and Black method [13], and levels of exchangeable Ca^{2+}, Mg^{2+}, K^+, and Na^+ (1 M NH_4-acetate pH 7.0, AA, Shimazu 660). The available P content was determined using the Lancaster method [14]. Heavy metals were extracted using the 0.1 M HCl solution and quantified using the ICPOES (inductively coupled plasma optical emission spectrophotometer, GBC model X-100, Australia).

Statistical analysis was performed with the SAS package, version 8.2. One-way ANOVA was carried out to compare the means of the different treatments where significant F values were detected.

RESULTS AND DISCUSSION

Evaluation of Reasonable Compost Application Level

The biomasses of the selected plants was significantly increased with increasing compost application rates up to 40 Mg·ha^{-1}, but thereafter sharply decreased. Similar growth trends were observed in all treatments as compared to the control, irrespective of the plant species and ET cover soil composition (Figure 2). Using a quadratic response model, the dry biomass yield of Cosmos bipinnatus in the composite ET cover that was made of bottom ash and soil was affected by the compost application rates as "Yield (kg·ha^{-1}) = 2234 + 94.7 Compost − 0.98 Compost^2 (model R^2= 0.756**)", where compost application rate is expressed as Mg·ha^{-1}. Using this equation, the maximum biomass yield was ca. 4522 kg·ha^{-1} at ca. 40 Mg·ha^{-1} compost application level, which is approximately two times higher than the biomass yield (ca. 2234 kg·ha^{-1}) in the control (no compost application). The other two plant species showed the maximum biomass yields at similar levels of compost application (ca. 40 Mg·ha^{-1}), and the dry biomass yield was increased by ca. 25% and ca. 45% in Lesoedeza cuneata and Lespedeza cyrtobotrya plants, respectively compared with the control. Almost similar plant growth responses were observed between the single soil ET cover and manure compost application.

In Korea, compost application at approximately 10 - 20 Mg·ha^{-1} is generally recommended for agricultural soils [15]. However, the fertility of alpine soil in this study was very low at an organic matter of 14.5 g·kg^{-1} and available phosphorus of 6 mg P·kg^{-1} contents (Table 1) relative to the 24 and 235 mg·kg^{-1} of the average organic matter and available phosphorus contents of a typical upland soil in Korea in the 1990s, respectively [15]. As a result, the highest dry biomass yield was observed in this ET cover system at higher compost application rate compared with the typical upland soil, irrespective of the ET cover soil composition. In general, the manure compost can act as effective surface mulch, increase the concentration of soil organic matter, improve tilth and water-holding capacity, suppress weeds, and provide a long-term supply of nutrients as the organic material decomposes [16,17]. For these reasons, compost application has been advocated as one component of sustainable agriculture [18,19].

However, the significantly improved biomass productivities of Lesoedeza cuneata and Lespedeza cyrtobotrya were observed in the composite ET cover of soil (65%) and bottom ash (35%) compared with those in the single soil ET covers. Since coal combustion ash has high content of plant available inorganic nutrients and alkaline pH, the beneficial effects of coal ash as a soil amendment is well known [21-27]. The addition of alkaline coal ash, which has a pH over 9.0 [20], can reduce soil acidity to a level suitable for agriculture [21] and can increase the availability of Si, Na, K, Ca, Mg, B, S and other trace nutrients [22-27]. The commercial use of coal ash as a fertilizer in crop production is uncommon in most countries, because coal ashes may also contain non-essential elements that adversely affect crop, soil and groundwater quality (e.g., As, B, Cd, Se) [28-30]. Despite potential negative effects on environmental quality, since coal continues to be the prime source of energy in Korea and contains high concentration of plant essential inorganic elements, the utilization of coal ash is likely to remain a serious issue.

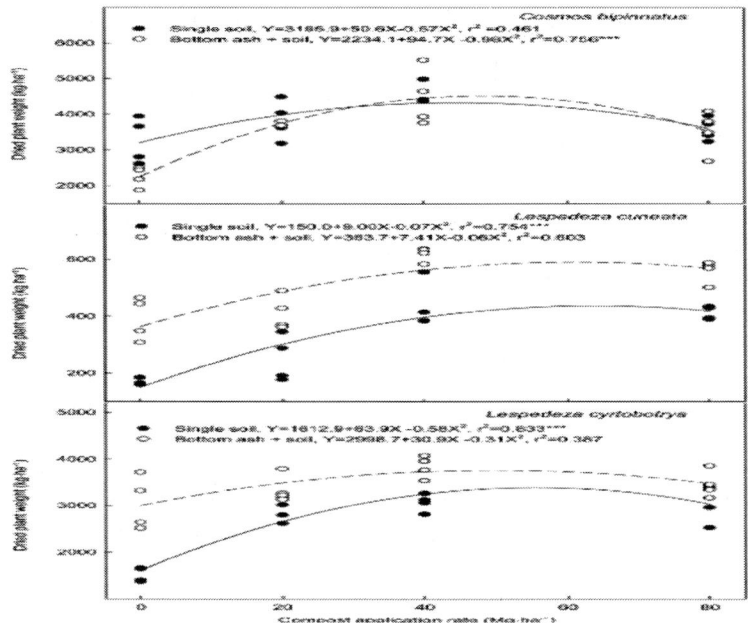

Figure 2: Changes of plant biomasses in the composite (soil and bottom ash mixture) and the single soil ET covers amended with different rates of manure compost in the pot test.

Effect of Bottom Ash and Compost on Re-Vegetation in the Vegetative Cover Pilot System

The vegetation compositions were changed in the landfill ET covers over the study years. Five different kinds of herbal grass and bush trees (Amorpha fruticosa, Artemisia princeps, Arundinella hirta, Lespedeza cuneata, Lespedeza cyrtobotrya) were sown in late June, 2007, and thereafter managed under the same condition for 3 years. However, green foxtail (Setaria viridis) which is a native plant in Korea was found as the dominant plant species in all treatments in the 1st year after the installation. Biomass productivity of the other sown plant species was very low probably due to late seeding. The system construction and stabilization was somewhat delayed and the plants were only sown in early summer season, not in spring. However, the proportion of the green foxtail to the total vegetation gradually declined over the 3 study years, but the growth of the sown plants significantly improved.

Total plant biomass productivity significantly increased with 40 $Mg \cdot ha^{-1}$ manure compost application, irrespective of the ET cover soil composition (Figure 3). In the 1st year, total plant biomass yield was ca. 1.27 and 1.43 $Mg \cdot ha^{-1}$ (on dry weight basis) in the single soil ET cover and the composite ET cover, respectively, but increased to ca. 1.73 and 1.94 times with the 40 $Mg \cdot ha^{-1}$ compost application. The effect of compost application on improving plant growth became clearer in the sterile soil ET cover compared with the high organic matter containing composite ET cover as time elapsed. In the 3rd

year after the installation, the total plant biomass was ca. 3.67 $Mg \cdot ha^{-1}$ on dry weight basis in the single soil ET cover, which increased to ca. 8.49 $Mg \cdot ha^{-1}$ with 40 $Mg \cdot ha^{-1}$ compost application. In comparison, the plant biomass productivity was not significantly different between 0 and 40 $Mg \cdot ha^{-1}$ compost application. This different response of plant growth characteristics with compost application in the 3rd year might have been caused by the difference of soil fertility between the two different ET cover soils. Among the soil chemical properties investigated at the plant harvesting stage in the 3rd year (2009) after the installation, soil fertility status such as pH, organic matter, and available inorganic nutrient contents were more favorable to plant growth in the composite ET layer than in the single soil layer (Table 1). In particular, the organic matter content of the composite ET covers was ca. 32 - 36 $g \cdot kg^{-1}$, which is much more than the organic matter content of 1.6 - 2.1 $g \cdot kg^{-1}$ in the single soil ET cover. The studied coal ash had ca. 33 $g \cdot kg^{-1}$ of organic matter. Therefore, coal bottom ash addition (35%) in the composite ET cover preparation significantly increased the organic matter content, and might have improved plant growth.

Soil organic matter is one of the most important constituents of soils due to its capacity in affecting plant growth indirectly and directly [31]. Indirectly, it improves the chemical and physical conditions of soils by increasing cation exchange capacity, buffering capacity, and enhancing aggregation, aeration and water retention. Improvement of soil biological properties affects soil microbial diversity and population, thereby creating a suitable environment for root growth of plants and soil microbes [32]. The most observable functions of soil organic matter include positive changes of soil physical properties such as bulk density, aggregate stability, porosity and water holding capacity when applied for long periods [33-34]. Generally, increased organic matter content of soils results in an increase in stability, irrespective of the origin of the stress [35].

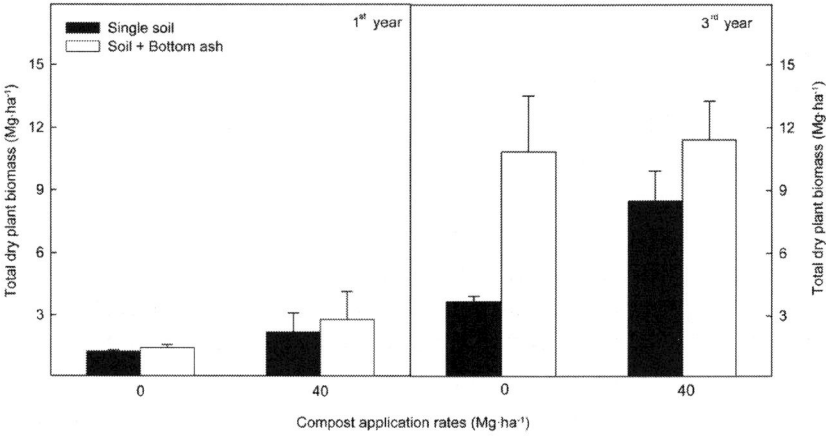

Figure 3: Plant biomass productivities in the composite (soil and bottom ash mixture) and the single soil ET covers amended with different rates of manure compost in the vegetative cover pilot test at the 1st and 3rd years after installation.

Plant growth was more significantly improved in the composite ET cover of soil and bottom ash compared to that in the single soil ET cover (Figure 3). The beneficial effect of the composite ET cover on improving plant biomass growth increased over the study years. Several studies have been conducted on use and disposal of coal by-products as soil amendments [29]. Bottom ash is a relatively coarse, gritty material in contrast to fly ash, which consists of very fine particles. As shown in Table 2, most components of soil fertility were significantly more favorable to plant growth in the composite ET cover soil than the single soil ET cover. As a result, the improvement of soil fertility might have become more effective in enhancing plant biomass growth in the composite ET cover.

There was a slight increase in the amounts of 0.1 M HCl-extractable heavy metals such as Cu, Pb and Zn, following the additions of bottom ash (Table 2). However, these values for heavy metals that were detected in this pilot experiment were lower than the criteria for soil pollution as regulated by the Korean government at 200 mg·kg^{-1}, 400 mg·kg^{-1} and 800 mg·kg^{-1} of Cu, Pb and Zn, respectively. Williams et al. [36] tested the land application of bark broiler bottom ash on moderately well drained Atlantic Coastal Plain soils and their findings revealed that the bottom ash application did not show any adverse effect of heavy metal pollution such as As, Cd, Cu, Cr, Ni on soil or ground water quality at the maximum application rates (44 Mg·ha^{-1}). As a result, our pilot system study indicates that bottom ash as a mixing material in the ET cover system has potential as a good soil additive that will not be detrimental to soil, plants, or the environment.

To conclude, manure compost was very effective to enhance the growth of three selected plants (Cosmos bipinnatus, Lespedeza cuneata, Lespedeza cyrtobotrya) in the composite ET cover system made of soil and bottom ash, and ca. 40 Mg·ha^{-1} of compost could be a reasonable application level in a sterile ET cover soil. Manure compost application significantly increased total plant biomass productivity, and might have stabilized the early vegetation development in the ET cover system. The effect of compost application on vegetative stabilization was more significantly improved in the composite ET cover of soil and bottom ash than that in the single soil ET cover, mainly due to a more favorable soil fertility conditions such as high content of organic matter, available inorganic elements, and neutral pH promoting plant growth. The bottom ash as a mixing material of soil in the ET cover system has a strong potential as an alternative to fine soil, and manure compost addition can effectively stimulate vegetative stabilization in the ET cover system.

Table 2: Chemical properties of ET vegetation media made of the composite of soil (65%) and coal bottom ash (35%), and the single soil collected in the vegetative cover pilot test at plant harvesting stage in the 3rd year after installation

ET cover Material	Single soil		Composite of BA and soil		LSD0.05
Compost applicatiom(Mg ha-1)	0	40	0	40	
pH(H2O, 1:5)	5.57	5.53	7.79	7.56	0.33

Elecrical conductivity(dS m-1)	0.14	0.23	0.37	0.36	0.09
Organic matter(g kg-1)	1.6	2.1	32.3	36.4	6.44
Available P(mg kg-1)	3.7	5.7	116.9	141.9	3.83
Exchasngeable cation(cmol+ kg-1)					
K	0.28	0.41	0.28	0.37	0.01
Ca	3.16	3.93	3.49	6.89	0.30
Mg	1.02	1.10	3.01	2.60	0.05
Na	0.11	0.11	0.46	0.21	0.01
0.1N HC1 extractable(mg kg-1)					
As	nd	nd	nd	nd	-
Cd	nd	nd	nd	nd	-
Cu	nd	1.06	4.39	4.16	0.19
cr	nd	nd	nd	nd	-
Ni	nd	nd	nd	nd	-
Pb	nd	nd	nd	1.71	0.46
Zn	0.57	1.72	3.13	8.47	0.38

Note: BA and nd mean bottom ash and not detected respectively.

REFERENCES

1.   US Environmental Protection Agency (1989) Final covers on hazardous waste landfill and surface impoundments.

2.   Jesionek, K.S., Dunn, R.J. and Daniel, D.E. (1995) Evaluation of landfill final covers. Proceedings of the 5th International Landfill Symposium, Sardinia, 2-6 October 1995, pp. 509-532.

3.   Lundgren, T (1995) Sluttäckning av avfallsupplag-krav, material, utförande, kontroll. Swedish Environmental Protection Agency, Sweden.

4.   Othman, M.A., Bonaparte, R., Gross, B.A. and Schmertmann, G.R. (1995) Design of MSW landfill cover systems. Geotechnical Special Publications, 53, 218-257.

5.   Elshorbagy, W.A. and Mohamed, A.M.O. (2000) Evaluation of using municipal solid waste compost in landfill closure caps in arid areas. Waste Management, 20, 499- 507.doi:10.1016/S0956-053X(00)00025-8.

6.   Mollamahmutoglu, M. and Yilmaz, Y. (2001) Potential use of fly ash and bentonite mixture as liner or cover at waste disposal areas. Environmental Geology, 40, 1316- 1324.doi:10.1007/s002540100355.

7.   Kim, S.O., Kim, P.J. and Yu, C. (2008) Evaluation on feasibility of industrial by-products for development of mono-layer landfill cover system. Journal of Korean Society of Environmental Engineers, 30, 1075-1086.

8.   Yun, S.W., Kang, S.I., Jin, H.G., Kim, P.J., Kim, S.O. and Yu. C. (2010) Evaluation on the effect of coal-ash as landfill cover material of mono-layer cover system through the field scale test. Geotechnical Engineering, 26, 81-91.

9.   Korcak, R.F. (1995) Utilization of coal combustion by-products in agriculture and horticulture. In: Karlen, D.L., Wright, R.J. and Kemper W.O., Eds., Agriculture Utilization of Urban and Industrial By-Products, ASACSSA-SSSA, Madison, 107-130.

10. Dobson, M.C. and Moffat, A.J. (1995) A re-evaluation of objections to tree planting on containment landfills. Waste Management and Research, 13, 579-600.

11. National Institute of Agricultural Sciences and Technology (2007) Water management in upper soil. National Institute of Agricultural Science and Technology, Korea Rural Development Administration, Suwon.

12. Korea Meterological Administration (2012) Monthly report of automatic weather system data from January to December. Korea Meteorological Administration, Seoul.

13. Allison, L.E. (1965) Organic carbon. In: Black C.A., Ed., Methods of Soil Analysis, ASA-CSSA-SSSA, Madison, 1367-1389.

14. Rural Development Administration (1988) Method of soil chemical analysis. National Institute of Agricultural Science and Technology, Suwon.

15.   Rural Development Administration (1999) Fertilization standard of crop. National Institute of Agricultural Science and Technology, Suwon.

16.   Ozores-Hampton, M. and Obreza, T.A. (1999) Composted waste use on Florida vegetable crops: A review. In: Warman, P.R. and Taylor, B.R., Eds., Proceedings of the International Composting Symposium, CBA Press, Truro, Halifax, 827-842.

17.   Evanylo, G.K. and Daniels, W.L. (1999) Paper mill sludge composting and compost utilization. Compost Science and Utilization, 7, 30-39.

18.   Arrouge, T., Moresoli, G. and Soucy, G. (1999) Primary and secondary sludge composting: A feasibility study. Pulp and Paper Canada, 100, 33-36.

19.   Edwards, L., Burney, J.R., Richter, G. and MacRae, A.H. (2000) Evaluation of compost and straw mulching on soil-loss characteristics in erosion plots of potatoes in Prince Edwards Island, Canada. Agriculture, Ecosystems and Environment, 81, 217-222.doi:10.1016/S0167-8809(00)00162-6.

20.   Cha, D.W., Lee, H.S. and Jung, J.H. (1999) Production and composition of the power plant coal ash in Korea. Proceedings of Agricultural Utilization of Fly Ash Symposium. Gyeongsang National University, Chinju, 1-23.

21.   Moliner, A.M. and Street, J.J. (1982) Effect of fly ash and lime on growth and composition of corn (Zea mays L.) on acid sandy soils. Soil and Crop Science Society of Florida Proceedings, 41, 217-220.

22.   Elseewi, A.A., Grimm, S.R., Page, A.L. and Straughan, I.R. (1981) Boron enrichment of plants and soils treated with coal ash. Journal of Plant Nutrition, 3, 409-427.doi:10.1080/01904168109362848.

23.   Druzina, V.D., Miroshrachenko, E.D. and Chertov, O.D. (1983) Effect of industrial pollution on nitrogen and ash content in meadow phytocoenotic plants. Botanicheskii Zhurnal, 68, 1583-1591.

24.   Wong, M.H. and Wong, J.W.C. (1989) Germination and seedling growth of vegetable crops in fly ash amended soils. Agriculture, Ecosystems and Environment, 26, 23- 35.doi:10.1016/0167-8809(89)90035-2.

25. Ko, B.G. (2000) Effects of fly ash and gypsum application on soil improvement and rice cultivation. Ph.D. Dissertation, Gyeongsang National University, Chinju.

26. Lee, H., Ha, H.S., Lee, C.H., Lee, Y.B. and Kim, P.J. (2006) Fly ash effect on improving soil properties and rice productivity in Korean paddy soils. Bioresource Technology, 97, 1490-1497. doi:10.1016/j.biortech.2005.06.020.

27. Lee, C.H., Lee, H., Lee, Y.B., Chang, H.H., Ali, M.A., Min, W., Kim, S. and Kim. P.J. (2007) Increase of available phosphorus by fly-ash application in paddy soils. Communications in Soil Science and Plant Analysis, 38, 1551-1562. doi:10.1080/00103620701378482.

28. Adriano, D.C., Woodford, T.A. and Ciravolo, T.G. (1978) Growth and elemental composition of corn and bean seedlings as influenced by soil application of coal ash. Journal of Environmental Quality, 7, 416-421.doi:10.2134/jeq1978.00472425000700030025x.

29. Adriano, D.C., Page, A.L., Elseewi, A.L., Chang, A.C. and Straughan, I. (1980) Utilization and disposal of fly ash and other coal residues in terrestrial ecosystems: A review. Journal of Environment Quality, 9, 333-344. doi:10.2134/jeq1980.00472425000900030001x.

30. Page, A. L., Elseewi, A.A. and Straughan, I.R. (1979) Physical and chemical properties of fly ash from coalfired power plants with reference to environmental impacts. Residue Reviews, 71, 83-20. doi:10.1007/978-1-4612-6185-8_2.

31. Bongiovanni, M.D. and Lobartini, J.C. (2006) Particulate organic matter, carbohydrate, humic acid contents in soil macroand microaggregates as affected by cultivation. Geoderma, 136, 660-665. doi:10.1016/j.geoderma.2006.05.002.

32. Senesi, N. and Loffredo, E. (1999) The chemistry of soil organic matter. In: Spark, D.L., Ed., Soil Physical Chemistry, CRC Press, Boca Raton, 239-370.

33. Edwards, C.A. and Lofty, J.R. (1982) Nitrogenous fertilizers and earthworms populations in agricultural soils. Soil Biology and Biochemistry, 147, 515-521. doi:10.1016/0038-0717(82)90112-2.

34. Schjonning, P. and Christensen, B.T. (1994) Physical and chemical properties of a sandy loam receiving animal manure, mineral fertilizer or no fertilizer for 90 years. European Journal of Soil Science, 45, 257-268. doi:10.1111/j.1365-2389.1994.tb00508.x.

35. Kay, B.D. (1998) Soil structure and organic carbon: A review. In: Lal, R., et al., Ed., Soil Processes and the Carbon Cycle, CRC Press, Boca Raton, 169-197.

36. Williams, T.M., Charles, A.H. and Smith, B.H. (1996) Forest soil and water chemistry following bark broiler bottom ash application. Journal of Environment Quality, 25, 955-961. doi:10.2134/jeq1996.00472425002500050005x.

Chapter 8

Use of Coal Waste as Fine Aggregates in Concrete Paving Blocks

Cassiano Rossi dos Santos, Juarez Ramos do Amaral Filho, Rejane Maria Candiota Tubino, and Ivo André Homrich Schneider

Federal University of Rio Grande do Sul—UFRGS, Post Graduate Programme in Mining, Metallurgical and Materials Engineering—PPGE3M, Porto Alegre, Brazil

ABSTRACT

The aim of this work was to study the use of coal waste to produce concrete paving blocks. The methodology considered the following steps: sampling of a coal mining waste; gravity separation of the fraction with specific gravity between 2.4 and 2.8; comminution of the material and particle size analysis; technological characterization of the material and the production of concrete paving blocks. The results showed that the coal waste considered in this work can be used to replace conventional sand as a fine aggregate for concrete paving blocks. This practice can collaborate in a cleaner coal production.

INTRODUCTION

The commercial coal production in the southern region of Brazil (comprising the Paraná, Santa Catarina, and Rio Grande do Sul states) has been occuring since the beginning of the twentieth century. Specifically in the Santa Catarina State, the production occurs at the "Irapuá", "Bonito", and, mainly, "Barro Branco" seams. These Gondwanic coals are classified for the major part as a highvolatile bituminous in rank. The thickness of the "Barro Branco" seam ranges from 1.66 to 2.27 m, with an average value of 1.80 m. However, net clean coal thickness is reduced to 0.47 - 1.48 m, due to the alternating layers of impure coal (shaley coal), carbonaceous shale, siltstone, and sandstones (Figure 1). Pyrite lenses, sometimes several centimeters thick, are also common [1].

Currently, the run-of-mine coal (ROM) is gravimetrically concentrated and almost entirely used for electricity generation. Due to the geological characteristics, large amounts of solid wastes are generated. It is estimated that more than 300 millions tonnes of coal waste exists in the south of Brazil, generating environmental impacts and economic costs. Regarding the Santa Catarina Coalfields, about 60% - 65% of the ROM coal is discharged at dump deposits as waste [2]. These wastes can lead to the formation of acid mine drainage (AMD), a source of ground and surface water pollution [3].

Through gravity concentration processes of this coal wastes it is possible to produce three output streams: 1) a low specific gravity material (relative density < 2.4) and more carbonaceous waste composed by shaley coal and carbonaceous shale; 2) an intermediated material (2.4 < relative density < 2.8) composed mainly by siltstone and sandstone; and 3) a high specific gravity material (relative density > 2.8) that is rich in pyrite. Presently, there has been some initiatives in Brazil to reprocess some coal wastes deposits with the purpose of recover part of the carbonaceous materials for energy production and, alternatively, to concentrate the pyrite for sulphuric acid production. However, still remains the intermediate density material, which represents 40% - 50% in mass of the coal waste deposit and could be considered as a geomaterial for possible use in civil construction and agriculture [2].

A further serious problem is that the productive chain of civil engineering uses huge amounts of raw materials. In recent years, rapid development has led to an increased demand for river sand, which is largely used as a fine aggregate for construction. The extraction of sand from river bed and river bank may cause adverse affects on the environment, like river bank erosion, river bed degradation, and deterioration of river water quality. This subject has been of concern in Brazil and other countries [4-6].

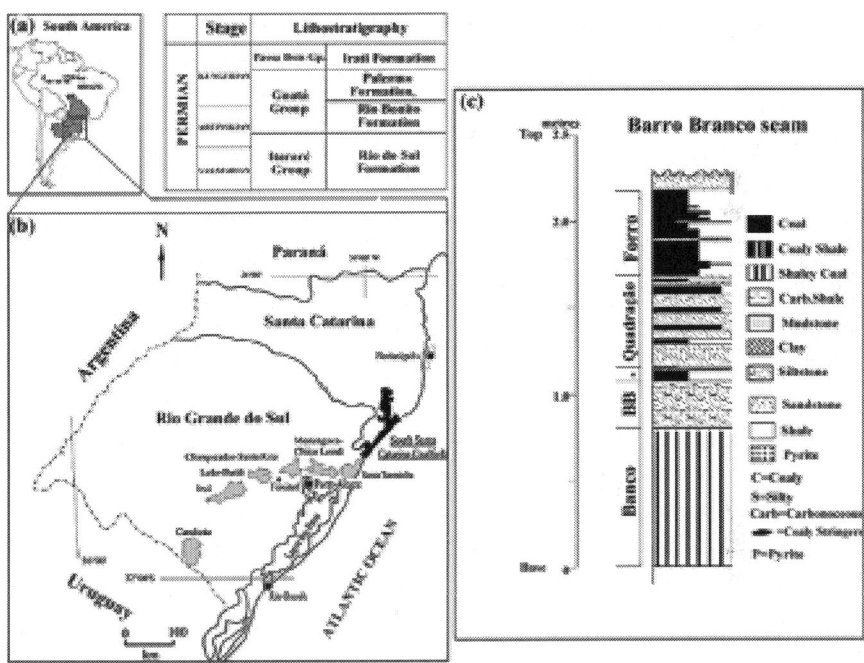

Figure 1: (a) Location of the southern Paraná Basin, Brazil; (b) Distribution of major coalfields in Rio Grande do Sul and Santa Catarina States; (c) Transect showing lithological profiles of the Barro Branco seam [1].

For example, segmented paving blocks are utilized worldwide and can be used in a large range of applications. The conventional source of fine aggregates for paving blocks are river sand or, alternatively, artificial sand by crushing rocks [7]. However, previous research has shown that it is also possible to use some wastes to produce concrete paving blocks: for example, gasification residues [8], crushed clay bricks [9], and ceramic tile production wastes [10].

Thus, the aim of this work was to study the use of coal waste as fine aggregates to produce concrete blocks for paving. The article briefly evaluates the main technical and environmental parameters that are involved and focused on the recycling part of the coal waste. The present study was undertaken from a scientific standpoint part of the effort to develop an effective coal cleaning process route focused on mitigating pollution problems associated with coal wastes worldwide.

EXPERIMENTAL

The coal waste was collected from the dump deposit of the "Verdinho Mine", Santa Catarina State-Brazil, which extracts the "Barro Branco" seam. The material was submitted to a laboratory dense medium separation processing that aimed at obtaining a fraction with relative density between 2.4 and 2.8 [11]. This fraction was crushed in a roller mill and sieved to reach a similar size particle distribution of river sand, which is commonly used for paving block production in Brazil. The quartz river sand was obtained from Jacuí River, Rio Grande do Sul State —Brazil. Technological characterization of both materials included particle size distribution, specific weight measurements, visual observation in a petrographic magnifying lens, and mineral phase determination by x-ray diffraction. Elemental analyses of the fine aggregates were carried out by x-ray fluorescence (for Si, Fe, Al, Ca, K) and high temperature decomposition in a CHNS analyzer (for C, H, N, and S). The main characteristics of both materials are resumed in Table 1.

The concrete paving blocks were produced in a vertical shaft concrete mixer. The reference trace used 5.36 kg of cement, 6.26 kg of coarse basaltic aggregate, 14.18 kg of river sand aggregate, and a water/cement ratio of 0.39. The coal waste was used as a substitute for river sand aggregate, considering the following volumetric levels of substitution: 0%, 25%, 50%, 75%, and 100%.

Table 1: Properties of the conventional quartz sand aggregate and the coal waste aggregate

Property	Conventional quartz sand aggregate	Coal waste aggregate
Particle size(mm)		
Size distribution	0.15 to 4.0	0.15 to 4.0
D90	2.0	3.0
DSO	0.4	1.0
DIO	0.15	0.15
Density (kg/dm³)		
Real	2.6	2.3
Apparent	1.6	1.3
Particle shape	Rounded and sub-rounded	Angular
Color	Yellowish	Grayish
Mineralogical composition	Quartz—SiO_2 (major mineral phase)	Quartz—SiO_2 major mineral phase), Kaolinite—$Al_4(OH)_8(Si_4O_{10})$, Illite-(K, H_3OXAl, Mg, Fe$)_2$(Si, Al$)_4O_{10}[(OH)_2$, $(H_2O)]$, Gypsum—$CaSO_4.2H_2O$.
Elemental composition (%)		
C	ND	2.6
H	ND	0.8
N	ND	0.1
S	ND	1.9
Si	63.7	47.0
Fe	1.0	7.5
Al	1.7	14.0
Mn	0.03	0.2
Ca	0.4	4.2
K	3.1	5.5

For each level of substitution, the water/cement ratio was reestablished to provide the same consistency of concrete [12]. The concrete pieces were molded in a manual press machine with a production capacity of six blocks per cycle. The blocks were molded in the "unipaver" shape with the following dimensions: 22.5 cm length, 12.0 cm width, and 8 cm height.

The technological characterization of the paving blocks included the resistance to compression (at 7, 28, and 90 days), abrasion resistance, water absorption, and colorimetric parameters (at 28 days) [13]. Compression strength, abrasion resistance, and water absorption were carried out in accordance, respectively, to Brazilian Standard Procedures NBR 9780 [14], NBR 12042 [15], and NBR 9778 [16]. Color measurements of the concrete blocks were recorded by reflectance spectra obtained using a MINOLTA CM-2600D spectrophotometer with an integrations sphere associated with an ultraviolet filter. The illuminant D65, which simulates daylight and the standard observer at $10°$ were chosen. At the beginning of the experiment, the calibration was done with two reference points, the zero and the white standard. The color parameters a^*, b^*, and L^* that corresponded to the uniform color space CIELAB were obtained directly from the apparatus [17]. Within the uniform space CIELAB, two color coordinates, a^* and b^*, as well as a psychometric index of lightness, L^*, are defined. a^* takes positive values for reddish colors and negative values for the greenish ones, whereas b^* takes positive values for yellowish colors and negative values for the bluish ones. L^* presents an approximate measurement of luminosity; according to this property each color can be considered as equivalent to a member of the grey scale, ranging between black and white, taking values within the range of 0 - 100. All measurements were carried out in 6 samples (n = 6), and the average and standard deviation was calculated. The statistical technique used to evaluate the differences between averages was the Analysis of Variance (ANOVA), considering a significance level of 5%.

Finally, the acid-generation potential of the raw waste, of the coal waste fine aggregates, and of the paving blocks at 28 days was conducted by the traditional method of accounting for acids and bases (ABA) [18]. The objective was to determine the balance between the minerals that produce acidity (acidity potential—AP) and the minerals which consume acidity (neutralization potential-NP). The net neutralization potential (NNP) was calculated from the difference between NP and AP. A sam- ple is classified as acid forming when it has NNP values less than -20 kg $CaCO_3$/t and as non-acid forming when it has NNP values greater than $+20$ kg $CaCO_3$/t. Samples are classified as uncertain when their values range from −20 to +20 kg $CaCO_3$/t.

Table 2: Properties of the concrete blocks for paving considering different levels of substitution of river sand by coal waste aggregate

Substitution of river sand aggregate by coal waste aggregate					
Property	0%	25%	50%	75%	100%
Water/cement ratio	0.35	0.37	0.39	0.43	0.44
Compressive strength (MPa) 7 days	28.1 ± 2.8[a]	33.0 ± 3A[b]	34.2 ± 1.3[b]	28.3 ± 1.6[a]	24.8 ± 4.6[a] 27.3 ± 3.1[b]
28 days	39.5 ± 2.9[a]	37.6 ±1.6[b]	36.6 ± 1.4[b]	31.2 ± 2.7[b]	27.2 ± 4.8[c]
90 days	40.7 ± 0.3[a]	36.2 ± 5.2[b]	34.1 ± 4A[b]	29.0 ± 3.8[c]	
Abrasion resistance (mm) 28 days	6.6 ± 0.0[a]	7.5 ± 1.3[a]	7.9 ± 0.0[a]	8.1 ± 0.5[a]	11.44 ± 3.0[b]
Water absorption (%) 28 days	4.9 ± 0.0[a]	5.3 ± 0.1[a]	5.4 ± 0.0[a]	6.910.6[b]	8.0 ±0.8[c]
Colorimetric properties 28 days L*	57.5 ± 2.7[a]	55.4 ± 5.3[a]	52.7 ± 4.1[a]	53.2 ± 3:3[a]	53.1 ± 3.0[a]
a*	0.1±0.2[a]	0.3 ± 0.5[b]	0.6 ± 0.3[b]	0.4 ± 0.2[b]	OA ± 0.1[b]
b*	7.4 ± 1.4[a]	6.2 ± 1.8[a]	7.6± 1.5[a]	7.8 ± 1.6[a]	6.4 ± 0.6[a]
Mass of fine aggregates of coal waste consumed per area of pavement (kg/m²)	0.0	12.1	24.2	36.4	48.5

Average ± standard deviation. Values with the same letters compared horizontally do not differ significantly from each other.

Table 3: Acid generation prediction results of the raw waste, coal waste fine aggregate and the concrete paving blocks with 0%, 25%, and 50% of substitution of river sand by coal waste aggregate

Parameter	Raw waste			Concrete paving blocks	
		Coal waste aggregate	0% Substitution	25% Substitution	50% Substitution
Total S (%)	7.0	1.9	0.5	0.4	0.9
AP (kg CaCO$_3$/t)	218.8	60.8	15.7	12.2	27.5
NP (kg CaCO$_3$/t)	0.0	0.0	241.0	430.0	488.2
NNP	—218.8	—60.8	225.3	417.8	460.7
Formation of AMD	Yes	Yes	No	No	No

RESULTS AND DISCUSSION

The mineral fine aggregates produced from the coal waste have quartz as their major crystalline phase. The presence of kaolin, illite, and gypsum was also detected. The particles are angular in shape, due to the rock fragmentation procedure. With regard to the presence of sulfur, the concentration in the fine aggregate was determined as 1.9%. This element is considered harmful for concretes, and most international standards recommend that the amount of sulfates and sulfides in aggregates for concrete production should not exceed the value of 1% [19]. Concerning the materials used in this work, the fine aggregates produced from coal tailings should be applied in levels of substitution of river sand of no more than 50%.

Table 2 shows the main technological properties of concrete blocks for paving, while considering the different levels of substitution. It can be observed that the concrete block produced with levels of substitution of 25% and 50%, at 28 days, statistically presents similar behavior of the reference blocks in terms of compressive resistance, abrasion resistance, and water absorption. In terms of the compression resistance, at 28 and 90 days the blocks attained, or are very close, to the value of 35 MPa which had been established by the Brazilian Legislation [13]. The colorimetric parameters demonstrate that the paving blocks produced with coal waste aggregates are imperceptibly darker (lower value of L*) when compared with the paving blocks produced with the conventional aggregate (Figure 2).

With regard to the acid generation (Table 3), the material collected from the coal waste deposit presents a high acid generation potential, with a sulfur content of 7.0%, AP of 218.8 kg $CaCO_3$/t, NP of 0.0 kg $CaCO_3$/t, and NNP of −218.8 kg $CaCO_3$/t. The fraction used for fine aggregate production, with a density between 2.4 and 2.8, exhibited a reduced acid generation potential, with a sulfur content of 1.9%, AP of 60.8 kg $CaCO_3$/t, NP of 0.0 kg $CaCO_3$/t, and NNP of −60.8 kg $CaCO_3$/t. The paving blocks produced with 25% of substitution of river sand by coal waste aggregate presented an S content of 0.4%, AP of 12.2 kg $CaCO_3$/t, NP of 430.0 kg $CaCO_3$/t, and NNP of 417.8 kg $CaCO_3$/t. These results showed that the manufacture of paving block provided an alkaline environment and prevented acid generation.

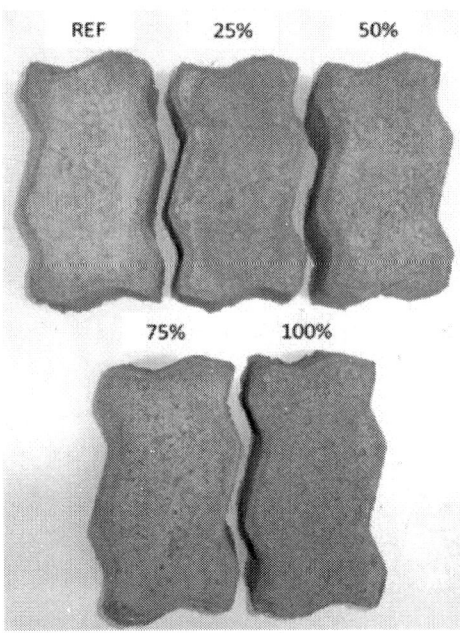

Figure 2: Concrete paving blocks produced with varied levels of substitution of river sand by coal waste aggregate.

With a safe level of substitution of 25%, the demand by coal waste aggregate was estimated to be 12.1 kg per m^2 of paved area. Considering a run-of-mine coal production of a typical Brazilian underground mine of 40,000 t per month, 60% of the material discharged as waste, and 50% of this fraction separated between the relative densities of 2.4 and 2.8, it is possible to produce about 12,000 t of fine aggregates, which is sufficient to attend a paved area of about 1 km^2.

Finally, it should be considered that many coal producers and coal consumer companies in the world are working to avoid the environmental impacts of the coal production chain [20]. The same effort has been carried out in the carboniferous region of Santa Catarina-Brazil, by means of the industries, government, and research institutions, which are engaged to recover all the pollution provided by decades of coal exploration. Many has been developed to recover the degraded areas [21] and to treat the acid mine drainage [3]. However, we consider that part of the solution is to provide a useful destination to the coal mining tailings [20,22,23], considering the principles of recycling, sustainable development, and integrated mine tailings management [20,24,25].

CONCLUSIONS

The results showed that it was possible to process the coal waste from the carboniferous region of Santa Catarina-Brazil—and obtain a recycled fine aggregate that can be used in civil construction. Concrete blocks for paving produced with 25% and 50% of recycled coal waste in substitution of river sand presented satisfactory results in terms of mechanical strength. The use of coal waste as a fine aggregate for concrete block paving manufacture presents technical viability and environmental benefits. This practice can collaborate in clean coal production and enable socioeconomic development within the regional context. The demand by sand deposits can be minimized and a part of coal tailings can be used, reducing the volume in coal waste deposits. We believe that this procedure can be applied to minimize the environmental problems in coal production in Brazil and other parts of the world.

ACKNOWLEDGEMENTS

The authors are grateful for the financial support extended by FINEP, CAPES, CNPq, and the Brazilian Coal Network for this research.

REFERENCES

1.   W. Kalkreuth, M. Holz, A, Mexias, M. Balbinot, J. Levandowski, J. Willett, R. Finkelman and H. Burger, "Depositional Setting, Petrology and Chemistry of Permian Coals from the Paraná Basin: 2. South Santa Catarina Coalfield, Brazil," International Journal of Coal Geology, Vol. 84, No. 3-4, 2010, pp. 213-236. doi:10.1016/j.coal.2010.08.008

2.   J. R. Amaral Filho, I. A. H. Schneider, R. M. C. Tubino, I. A. S. Brum, G. Miltzarek, C. H. Sampaio and C. H. Schneider, "Characterization of a Coal Tailing Deposit for Zero Waste Mine in the Brazilian Coal Field of Santa Catarina," In: Ch. Wolkersdorfer and A. Freund, Eds., Mine Water & Innovative Thinking, CBU Press, Sydney, 2010, pp. 639-643.

3.   R. D. R. Silva and J. Rubio, "Treatment of Acid Mine Drainage (AMD) from Coal Mines in South Brazil," International Journal of Coal Preparation and Utilization, Vol. 29, No. 4, 2009, pp. 192-202. doi:10.1080/19392690903066045

4.   F. I. Santo and L. E. Sánchez, "GIS Applied to Determine Environmental Impacts Indicators Made by Sand Mining in Floodplains in Southeasterrn Brazil," Environmental Geology, Vol. 41, No. 6, 2002, pp. 628-637. doi:10.1007/s002540100441

5.   M. Sandecki, "Aggregate Mining in River Systems," California Geology, Vol. 42, No. 4, 1989, pp. 88-94.

6.   Ministry of Natural Resources and Environment, Department of Irrigation and Drainage Malaysia, "River Sand Mining Management Guideline," Kuala Lumpur, 2009, p. 85.

7.   American Concrete Institute—ACI, "Aggregates for Concrete," Education Bulletin E1-07, Farmington Hills, 2007, p. 29.

8.   E. Holt and P. Raivio, "Use of Gasification Residues in Compacted Concrete Paving Blocks," Cement and Concrete Research, Vol. 36, No. 3, 2006. pp. 441-448.doi:10.1016/j.cemconres.2005.09.013

9. C. S. Poon and D. Chan, "Paving Blocks Made with Recycled Concrete Aggregate and Crushed Clay Brick," Construction and Building Materials, Vol. 20, No. 8, pp. 569-577.doi:10.1016/j.conbuildmat.2005.01.044

10. S. D. Wattanasiriwech, A. Saiton and S. Wattanasiriwech, "Paving Blocks from Ceramic Tile Production Waste," Journal of Cleaner Production, Vol. 17, No. 18, 2009, pp. 1663-1668. doi:10.1016/j.jclepro.2009.08.008

11. J. W. Leonard, "Coal Preparation," 5th Edition, Society for Mining, Metallurgy, and Exploration, Inc., Littleton, 1991.

12. P. K. Mehta and P. J. M. Monteiro, "Concrete: Structure, Properties, and Materials," 2nd Edition, Prentice-Hall, Englewood Cliffs, 1993.

13. ABNT—Associação Brasileira de Normas Técnicas, NBR 9781, "Peças de Concreto Para Pavimentação—Especificação," Rio de Janeiro, 1987.

14. ABNT—Associação Brasileira de Normas Técnicas, NBR 9780, "Peças de Concreto Para Pavimentação. Determinação da Resistência à Compressão—Método de Ensaio," Rio de Janeiro, 1987.

15. ABNT—Associação Brasileira de Normas Técnicas, NBR 12042, "Materiais Inorgânicos—Determinação do Desgaste por Abrasão—Método de Ensaio," Rio de Janeiro, 1992.

16. ABNT—Associação Brasileira de Normas Técnicas, NBR 9778, "Argamassa e Concreto Endurecidos—Determina- ção da Absorção de Água, Índice de Vazios e Massa Específica—Método de Ensaio," Rio de Janeiro, 2005.

17. CIE, "Recommendations on Uniform Color Spaces, ColorDifference Equations, Psychometric Color Terms," Bureau Central de la CIE, Viena, 1978.

18. Environmental Protection Agency, "Acid Mine Drainage Prediction," EPA 530-R-94-036, Technical Document, 1994.

19. C. H. Mattus and T. M. Gilliam, "A Literature Review of Mixed Waste Components: Sensitives and Effects upon Solidification/Stabilization in Cement-based Matrices," Oak Ridge National Laboratory, Martin Marietta Energy Systems, Inc., 1994.

20. L. Haibin and L. Zhenling, "Recycling Utilization Patterns of Coal Mining Waste in China," Resources, Conservation and Recycling, Vol. 54, No. 12, 2010, pp. 1331-1340.doi:10.1016/j.resconrec.2010.05.005

21. G. Milioli, "Mining, Environment, and Development in South Santa Catarina, Brazil. Non-Governmental Organization 'Terra Verde' and Its Ideas for Sustainability," Environments, Vol. 33, No. 1, 2005, pp. 25-40.

22. K. M. Skarzynska, "Reuse of Coal Mining Wastes in Civil Engineering—Part 2: Utilization of Minestone," Waste Management,Vol.15,No.2,1995,pp.83-126.doi:10.1016/0956-053X(95)00008-N

23. V. G. Lemeshev, I. K. Gubin, Y. A. Savel´ev, D. V. Tumanov and D. O. Lemeshev, "Utilization of Coal-Mining Waste in the Production of Building Ceramic Materials," Glass and Ceramics, Vol. 61, No. 9-10, 2004, pp. 308- 311.doi:10.1023/B:GLAC.0000048698.58664.97

24. M. Benzaazoua, B. Bussière, I. Demers, M. Aubertin, E. Fried and A. Blier, "Integrated Mine Tailings Management by Combining Environmental Desulphurization and Cemented Paste Backfill: Application to Mine Doyon, Quebec, Canada," Minerals Engineering, Vol. 21, No. 4, 2008, pp. 330-340. doi:10.1016/j.mineng.2007.11.012

25. A. H. Hesketh, J. L. Broadhurst and S. T. L. Harrison, "Mitigating the Generation of Acid Mine Drainage from Copper Sulfide Tailings Impoundments in Perpetuity: A Case Study for an Integrated Management Strategy," Minerals Engineering, Vol. 23, No. 3, 2010, pp. 225-229. doi:10.1016/j.mineng.2009.09.020

Citations

CHAPTER 1

S. Sivakumar and B. Kameshwari, "Influence of Fly Ash, Bottom Ash, and Light Expanded Clay Aggregate on Concrete," Advances in Materials Science and Engineering, vol. 2015, Article ID 849274, 9 pages, 2015. doi:10.1155/2015/849274.

CHAPTER 2

Sujata Mandal and Bhaskar D. Kulkarni, "Separation Strategies for Processing of Dilute Liquid Streams," International Journal of Chemical Engineering, vol. 2011, Article ID 659012, 19 pages, 2011. doi:10.1155/2011/659012.

CHAPTER 3

Ahmed A. S. Seifelnassr and Abdel-Zaher M. Abouzeid, "Exploitation of Bacterial Activities in Mineral Industry and Environmental Preservation: An Overview," Journal of Mining, vol. 2013, Article ID 507168, 13 pages, 2013. doi:10.1155/2013/507168.

CHAPTER 4

Kevin Lombard, Mick O'Neill, April Ulery, et al., "Fly Ash and Composted Biosolids as a Source of Fe for Hybrid Poplar: A Greenhouse Study," Applied and Environmental Soil Science, vol. 2011, Article ID 475185, 11 pages, 2011. doi:10.1155/2011/475185.

CHAPTER 5

Ruiqiang Liu and Rattan Lal, "Nanoenhanced Materials for Reclamation of Mine Lands and Other Degraded Soils: A Review," Journal of Nanotechnology, vol. 2012, Article ID 461468, 18 pages, 2012. doi:10.1155/2012/461468.

CHAPTER 6

N. Zarlin, T. Sasaoka, H. Shimada and K. Matsui, "Numerical Study on an Applicable Underground Mining Method for Soft Extra-Thick Coal Seams in Thailand," Engineering, Vol. 4 No. 11, 2012, pp. 739-745. doi: 10.4236/eng.2012.411095.

CHAPTER 7

S. Lee, S. Kim, C. Yu, S. Kim and P. Kim, "Vegetation of mono-layer landfill cover made of coal bottom ash and soil by compost application," Journal of Agricultural Chemistry and Environment, Vol. 2 No. 3, 2013, pp. 50-58. doi:10.4236/jacen.2013.23008.

CHAPTER 8

C. Santos, J. Filho, R. Tubino and I. Schneider, "Use of Coal Waste as Fine Aggregates in Concrete Paving Blocks," Geomaterials, Vol. 3 No. 2, 2013, pp. 54-59. doi: 10.4236/gm.2013.32007.

Index